Date: 3/29/18

J 509.22 IGN
Ignotofsky, Rachel,
Women in science : 50
fearless pioneers who

**PALM BEACH COUNTY
LIBRARY SYSTEM**
3650 Summit Boulevard
West Palm Beach, FL 33406-4198

D1682424

Women in Science

WOMEN IN SCIENCE

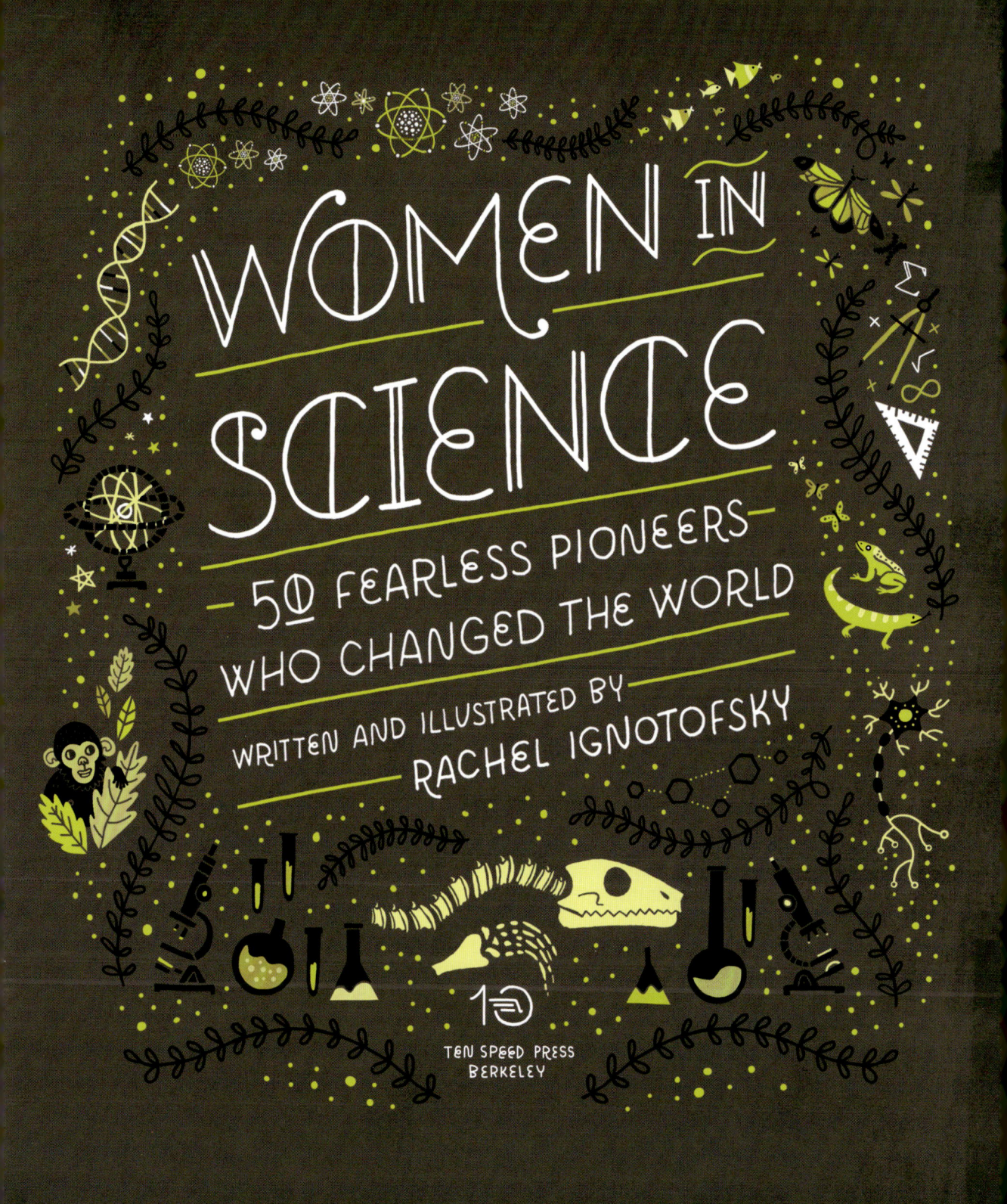

50 FEARLESS PIONEERS WHO CHANGED THE WORLD

WRITTEN AND ILLUSTRATED BY RACHEL IGNOTOFSKY

Ten Speed Press
Berkeley

CONTENTS

INTRODUCTION 6
HYPATIA (350-370CE — 415CE [?]) 9
MARIA SIBYLLA MERIAN (1647—1717) 11
WANG ZHENYI (1768—1797) 13
MARY ANNING (1799—1847) 15
ADA LOVELACE (1815—1852) 17
ELIZABETH BLACKWELL (1821—1910) 19
HERTHA AYRTON (1854—1923) 21
KAREN HORNEY (1885—1952) 23
NETTIE STEVENS (1861—1912) 25
FLORENCE BASCOM (1862—1945) 27
MARIE CURIE (1867—1934) 29
MARY AGNES CHASE (1869—1963) 31
TIMELINE 32
LISE MEITNER (1878—1968) 35
LILLIAN GILBRETH (1878—1972) 37
EMMY NOETHER (1882—1935) 39
EDITH CLARKE (1883—1959) 41
MARJORY STONEMAN DOUGLAS (1890—1998) 43
ALICE BALL (1892—1916) 45
GERTY CORI (1896—1957) 47
JOAN BEAUCHAMP PROCTER (1897—1931) 49
CECILIA PAYNE-GAPOSCHKIN (1900—1979) 51
BARBARA MCCLINTOCK (1902—1992) 53
MARIA GOEPPERT-MAYER (1906—1972) 55
GRACE HOPPER (1906—1992) 57
RACHEL CARSON (1907—1964) 59
LAB TOOLS 60
RITA LEVI-MONTALCINI (1909—2012) 63
DOROTHY HODGKIN (1910—1994) 65

CHIEN-SHIUNG WU (1912–1997)	67
HEDY LAMARR (1914–2000)	69
MAMIE PHIPPS CLARK (1917–1983)	71
GERTRUDE ELION (1918–1999)	73
KATHERINE JOHNSON (1918–)	75
JANE COOKE WRIGHT (1919–2013)	77
ROSALIND FRANKLIN (1920–1958)	79
ROSALYN YALOW (1921–2011)	81
ESTHER LEDERBERG (1922–2006)	83
STATISTICS IN STEM	84
VERA RUBIN (1928–2016)	87
ANNIE EASLEY (1933–2011)	89
JANE GOODALL (1934–)	91
SYLVIA EARLE (1935–)	93
VALENTINA TERESHKOVA (1937–)	95
PATRICIA BATH (1942–)	97
CHRISTIANE NÜSSLEIN-VOLHARD (1942–)	99
JOCELYN BELL BURNELL (1943–)	101
SAU LAN WU (194?–)	103
ELIZABETH BLACKBURN (1948–)	105
KATIA KRAFFT (1942–1991)	107
MAE JEMISON (1956–)	109
MAY-BRITT MOSER (1963–)	111
MARYAM MIRZAKHANI (1977–)	113
MORE WOMEN IN SCIENCE	114
CONCLUSION	117
GLOSSARY	118
SOURCES	122
ACKNOWLEDGMENTS	124
ABOUT THE AUTHOR	125
INDEX	126

INTRODUCTION

Nothing says trouble like a woman in pants. That was the attitude in the 1930s, anyway; when Barbara McClintock wore slacks at the University of Missouri, it was considered scandalous. Even worse, she was feisty, direct, incredibly smart, and twice as sharp as most of her male colleagues. She did things her way to get the best results, even if it meant working late with her students, who were breaking curfew. If you think these seem like good qualities for a scientist, then you are right. But back then, these weren't necessarily considered good qualities in a woman. Her intelligence, her self-confidence, her willingness to break rules, and of course her pants were all considered shocking!

Barbara had already made her mark on the field of genetics with her groundbreaking work at Cornell University, mapping chromosomes using corn. This work is still important in scientific history. Yet while working at the University of Missouri, Barbara was seen as bold and unladylike. The faculty excluded her from meetings and gave her little support with her research. When she found out that they would fire her if she got married and there was no possibility of promotion, she decided she had had enough.

Risking her entire career, she packed her bags. With no plan, except an unwillingness to compromise her worth, Barbara went off to find her dream job. This decision would allow her to joyously research all day and eventually make the discovery of jumping genes. This discovery would win her a Nobel Prize and forever change how we view genetics.

Barbara McClintock's story is not unique. As long as humanity has asked questions about our world, men and women have looked to the stars, under rocks, and through microscopes to find the answers. Although both men and women have the same thirst for knowledge, women have not always been given the same opportunities to explore the answers.

In the past, restrictions on women's access to education were not uncommon. Women were often not allowed to publish scientific papers. Women were expected to grow up to exclusively become good wives and mothers while their husbands provided for them. Many people thought women were just not as smart as men. The women in this book had to fight these stereotypes to have the careers they wanted. They broke rules, published under pseudonyms, and worked for the love of learning alone. When others doubted their abilities, they had to believe in themselves.

When women finally began gaining wider access to higher education, there was usually a catch. Often they would be given no space to work, no funding, and no recognition. Not allowed to enter the university building because of her gender, Lise Meitner did her radiochemistry experiments in a dank basement. Without funding for a lab, physicist and chemist Marie Curie handled dangerous radioactive elements in a tiny, dusty shed. After making one of the most important discoveries in the history of astronomy, Cecilia Payne-Gaposchkin still got little recognition, and for decades her gender limited her to work as a technical assistant. Creativity, persistence, and a love of discovery were the greatest tools these women had.

Marie Curie is now a household name, but throughout history there have been many other great and important women in the fields of science, technology, engineering, and mathematics (STEM). Many did not receive the recognition they deserved at the time and were forgotten. When thinking of physics, we should name not only Albert Einstein but also the genius mathematician Emmy Noether. We should all know that it was Rosalind Franklin who discovered the double helix structure of DNA, not James Watson and Francis Crick. While admiring the advances in computer technology, let us remember not only Steve Jobs or Bill Gates, but also Grace Hopper, the creator of modern programming.

Throughout history many women have risked everything in the name of science. This book tells the stories of some of these scientists, from ancient Greece to the modern day, who in the face of "No" said, "Try and stop me."

ONE OF THE FIRST RECORDED WOMEN TO STUDY & TEACH MATH.

HAS BECOME A SYMBOL FOR ENLIGHTENMENT AND FEMINISM.

AN EXPERT IN PHILOSOPHY, ASTRONOMY, AND MATHEMATICS.

"IN SPEECH ARTICULATE AND LOGICAL, IN HER ACTIONS PRUDENT AND PUBLIC-SPIRITED... THE CITY GAVE HER SUITABLE WELCOME AND ACCORDED HER SPECIAL RESPECT." —THE *SUDA* LEXICON

HYPATIA

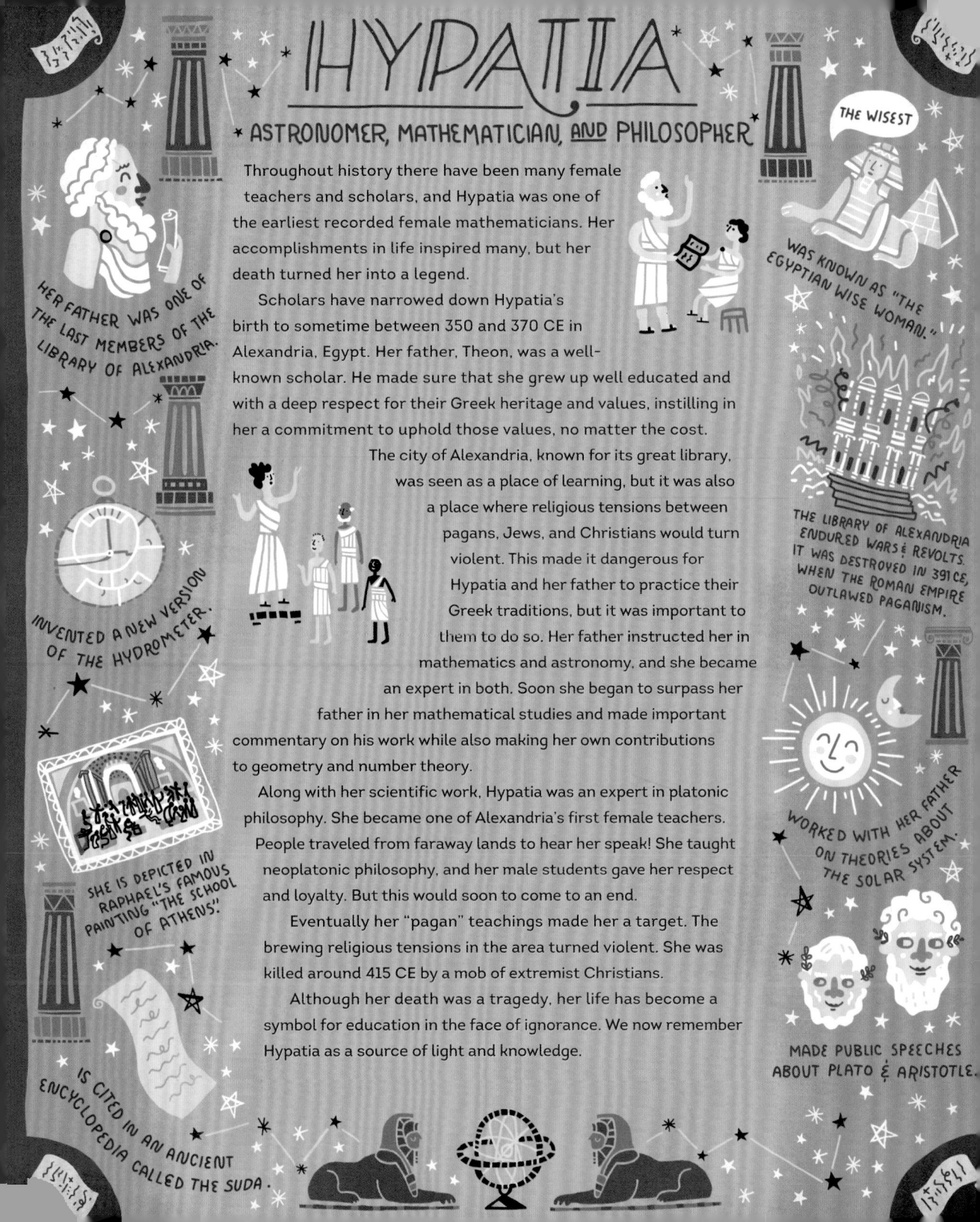

ASTRONOMER, MATHEMATICIAN, AND PHILOSOPHER

Throughout history there have been many female teachers and scholars, and Hypatia was one of the earliest recorded female mathematicians. Her accomplishments in life inspired many, but her death turned her into a legend.

Scholars have narrowed down Hypatia's birth to sometime between 350 and 370 CE in Alexandria, Egypt. Her father, Theon, was a well-known scholar. He made sure that she grew up well educated and with a deep respect for their Greek heritage and values, instilling in her a commitment to uphold those values, no matter the cost.

The city of Alexandria, known for its great library, was seen as a place of learning, but it was also a place where religious tensions between pagans, Jews, and Christians would turn violent. This made it dangerous for Hypatia and her father to practice their Greek traditions, but it was important to them to do so. Her father instructed her in mathematics and astronomy, and she became an expert in both. Soon she began to surpass her father in her mathematical studies and made important commentary on his work while also making her own contributions to geometry and number theory.

Along with her scientific work, Hypatia was an expert in platonic philosophy. She became one of Alexandria's first female teachers. People traveled from faraway lands to hear her speak! She taught neoplatonic philosophy, and her male students gave her respect and loyalty. But this would soon to come to an end.

Eventually her "pagan" teachings made her a target. The brewing religious tensions in the area turned violent. She was killed around 415 CE by a mob of extremist Christians.

Although her death was a tragedy, her life has become a symbol for education in the face of ignorance. We now remember Hypatia as a source of light and knowledge.

HER FATHER WAS ONE OF THE LAST MEMBERS OF THE LIBRARY OF ALEXANDRIA.

INVENTED A NEW VERSION OF THE HYDROMETER.

SHE IS DEPICTED IN RAPHAEL'S FAMOUS PAINTING "THE SCHOOL OF ATHENS".

IS CITED IN AN ANCIENT ENCYCLOPEDIA CALLED THE SUDA.

THE WISEST

WAS KNOWN AS "THE EGYPTIAN WISE WOMAN."

THE LIBRARY OF ALEXANDRIA ENDURED WARS & REVOLTS. IT WAS DESTROYED IN 391 CE, WHEN THE ROMAN EMPIRE OUTLAWED PAGANISM.

WORKED WITH HER FATHER ON THEORIES ABOUT THE SOLAR SYSTEM.

MADE PUBLIC SPEECHES ABOUT PLATO & ARISTOTLE.

ONE OF THE FIRST AND MOST IMPORTANT ENTOMOLOGISTS.

CLASSIFIED MANY NEW INSECT SPECIES.

CAREFULLY ILLUSTRATED THE METAMORPHOSIS OF THE BUTTERFLY.

"ART AND NATURE SHALL ALWAYS BE WRESTLING UNTIL THEY EVENTUALLY CONQUER ONE ANOTHER SO THAT THE VICTORY IS THE SAME STROKE AND LINE." — MARIA SIBYLLA MERIAN

MARIA SIBYLLA MERIAN
SCIENTIFIC ILLUSTRATOR AND ENTOMOLOGIST

Born in Germany in 1647, Maria Sibylla Merian combined science and art to become one of the greatest scientific illustrators of all time.

In the 1600s, Europeans did not have a basic understanding of insects. Most people thought that they were simply disgusting and not worth careful study. Maria could not have disagreed more. At a young age she started collecting insects to study how they behaved. Her stepfather taught her how to use paint, which she used to illustrate the different stages of her favorite insects' lives.

Maria was particularly interested in butterflies. At the time, no one really understood the connection between caterpillars and butterflies. In 1679, she published a book on metamorphosis, filled with scientific notes and illustrations.

Then Maria's life changed drastically. She left her husband and took her mother and 2 daughters to Holland. They joined a strict religious group that had ties with a Dutch colony in South America called Suriname. The mismanaged religious group fell apart, but Maria's interest in Suriname stayed with her.

Curious about new insects, at the age of 52, Maria braved the rainforests of South America. She documented never-before-seen bugs in the face of dangerous rain and heat. Unfortunately, her trip ended early when she contracted malaria, but she had already made the illustrations she needed to create her greatest book. *The Metamorphosis of the Insects of Suriname* was published in 1705 and became a hit all over Europe!

Maria's work helped future scientists to classify and understand insects, and her beautiful, detailed illustrations amaze and educate people to this day.

PEOPLE THOUGHT MARIA LOVED BUGS BECAUSE HER MOM VISITED AN INSECT COLLECTION WHILE PREGNANT.

PEOPLE USED TO CALL INSECTS "THE BEASTS OF THE DEVIL."

MARIA OBSERVED & PAINTED LIVE INSECTS WHILE OTHERS OBSERVED ONLY DEAD ONES IN DISPLAY CASES.

PEOPLE USED TO THINK INSECTS WOULD SPONTANEOUSLY APPEAR IN GARBAGE LIKE MAGIC.

MARIA'S PORTRAIT WAS ON GERMAN MONEY AND STAMPS.

COCOONS WERE ONCE CALLED "DATE PITS" IN GERMANY.

SHE HANDLED POISONOUS BUGS IN THE RAINFOREST.

WROTE POLITICAL POETRY ABOUT INJUSTICE.

WROTE PAPERS EXPLAINING TRIGONOMETRY AND THE PRINCIPLES OF MULTIPLICATION AND DIVISION.

ACCURATELY RECORDED LUNAR ECLIPSES & EQUINOXES.

"IT'S MADE TO BELIEVE / WOMEN ARE THE SAME AS MEN; / ARE YOU NOT CONVINCED / DAUGHTERS CAN ALSO BE HEROIC?" — WANG ZHENYI'S POETRY

WANG ZHENYI

ASTRONOMER, POET, AND MATHEMATICIAN

Wang Zhenyi was one of the greatest scholars in China. She was born in 1768, during the Qing dynasty. At the time, China had a strict feudal system; education was available only for the wealthy, and women were expected to cook, sew, and not be "bothered" by studies.

Wang Zhenyi was fortunate to be born into a family of scholars who valued her education. Her grandfather and father taught her astronomy and math. She also traveled extensively and saw how extreme taxation affected the less fortunate. Learning about the harshness of poverty inspired her to write poetry decrying injustice.

In Wang Zhenyi's day, eclipses were considered mysterious and beautiful but were not well understood. But she had theories about how they worked, and she created her own eclipse model using a mirror, a lamp, and a globe that she tied up with ropes around a table. She used it to prove her theories about how the moon blocks our view of the sun—or the earth blocks the sun's light from reaching the moon—during an eclipse.

And there were more planetary problems to solve! Wang Zhenyi scientifically studied the Chinese calendar system and used her telescope to measure the stars and further explain the rotation of the solar system.

She was also a dedicated mathematician. Her struggles with math would often make her stop and sigh, but she pushed through those tough moments. She understood complicated arithmetic theories and at the age of 24 published a 5-volume guide for beginners, called *Simple Principles of Calculation*. This work, compiled 6 years after Wang Zhenyi's death, was prefaced by the famous scholar Qian Yiji and read by many.

Wang Zhenyi lived only to age 29, but she is remembered as one of the greatest minds of the Qing dynasty. She published many volumes of writing on math, astronomy, and poetry, and her work influenced legions of scientists, mathematicians, and writers who came after her.

LOVED HER GRANDFATHER'S HUGE LIBRARY OF BOOKS.

UNDERSTOOD THAT THE EARTH WAS ROUND AND DESCRIBED IT AS A BALL.

LEARNED FROM WESTERN & EASTERN CALENDARS.

WAS ACCOMPLISHED IN ARCHERY & HORSEBACK RIDING.

EXPLAINED ECLIPSES IN HER PAPER "THE DISPUTE OF THE PROCESSION OF THE EQUINOXES".

UPDATED THE COUNT AND PLACEMENT OF THE STARS.

DEVELOPED HER OWN ARGUMENTS ON GRAVITY.

WROTE COMMENTARIES ON THE PYTHAGOREAN THEOREM AND OTHER TRIGONOMETRIC STUDIES.

$a^2 + b^2 = c^2$

HER WORK CHANGED HOW WE UNDERSTAND PREHISTORIC LIFE.

SHE FOUND THE FIRST SKELETONS OF THE ICHTHYOSAUR & THE PLESIOSAUR.

HER WORK WAS KEY IN PROVING THAT EXTINCTION OCCURS.

"THE GREATEST FOSSILIST THE WORLD EVER KNEW." — THE BRITISH JOURNAL OF THE HISTORY OF SCIENCE

MARY ANNING
FOSSIL COLLECTOR AND PALEONTOLOGIST

Mary Anning was born in 1799 in a small English seaside town called Lyme Regis. Her family was very poor, so to make ends meet she would help her father collect fossils to sell to rich tourists. It was dangerous work; the cliffs were steep, and the ocean caused riptides and landslides. Despite this, 11-year-old Mary took over the fossil business when her father died.

There was a time when people had never heard of dinosaurs and thought it was impossible for any animal species to go extinct. Mary helped to prove this wrong, and her discoveries began at a young age. When she was around 12, she discovered the first complete ichthyosaur skeleton ever found. She went on to discover 2 skeletons of the previously unknown species called plesiosaur. These fossils were unlike any animal known to humans, proving that extinction can occur!

She also discovered the first pterosaur skeleton outside of Germany and many different ancient fossilized fish. She helped determine that mysterious stones called bezoars were actually fossilized poop! Studying dinosaur poop is important to figuring out how they lived.

Despite her scientific accomplishments, she was not allowed to publish because she was a woman. Doctors and geologists respected her ideas and used her findings in their own work. Her name would be edited out, or never included to begin with. Although this was unfair, it was remarkable in Victorian England that she, a working class woman, was even allowed to mingle with educated gentlemen.

Mary Anning's discoveries allowed the world to see fossils as more than mystical oddities and introduced us to the age of the reptiles.

Her dog, Tray, came with her on fossil digs until he died in a landslide.

There is a myth that her genius came from being struck by lightning as a child.

It's rumored that the tongue-twister "she sells sea shells" is about Mary Anning.

She sold fossils to noble gentlemen.

Her life inspired many modern fictional stories.

People called fossils "devils toenails" and "snake stones".

WAS THE FIRST PERSON TO CREATE A COMPUTER PROGRAM.

WROTE ONE OF THE MOST IMPORTANT DOCUMENTS IN COMPUTER HISTORY.

IS HONORED WITH ADA LOVELACE DAY.

"IMAGINATION IS THE DISCOVERING FACULTY, PRE-EMINENTLY. IT IS THAT WHICH PENETRATES INTO THE UNSEEN WORLDS AROUND US, THE WORLDS OF SCIENCE." —ADA LOVELACE

ADA LOVELACE

MATHEMATICIAN AND WRITER

When Ada Lovelace first saw the Difference Engine, she became obsessed. The early computing pioneer Charles Babbage invented this gigantic, gear-filled calculator, and after meeting him in 1833, Ada did everything she could to convince him to work with her.

Ada's love affair with math started when she was very young. Her mother, Anne Isabella Milbanke, nicknamed the "Princess of Parallelograms," was a mathematician who wanted the right upbringing for her daughter. Ada's father was the famed poet Lord Byron. The wildness that made him an amazing poet also made Byron something of a lousy husband, which led Ada's mother to leave him after Ada was born. Her mother gave Ada an unusually strict mathematical education.

Ada met Charles Babbage when she was 17 and a very persistent young woman. She begged him to take her on as a student, but he was much too busy thinking up his next mechanical breakthrough. So when Ada saw an article in a Swiss journal about his newest idea, the Analytical Engine, she saw her chance to impress him.

The article was written in French, which Ada spoke, so she translated his paper into English and published it in 1843. But that wasn't all: she added her own notes, making it twice as long! This got Charles's attention, and their collaboration began.

Ada imagined a world where computers did more than mere calculations—a world where they could write music and become extensions of human thought. She also designed a way to program the Analytical Engine, using punch cards with a stepwise sequence of rational numbers called Bernoulli numbers. This is recognized as the first computer program ever!

Ada was a true visionary, and she remains an inspiration to this day. Her name has become a call to action and proof that women can accomplish great things in technology, computing, and programming.

She described herself as a poetical scientist.

Her last name comes from her husband, William King, the Earl of Lovelace.

Ada Lovelace Day is celebrated on the 2nd Tuesday in October.

She has inspired characters in narrative & graphic novels.

The US Department of Defense named a computer language "Ada".

She signed her letters to Charles Babbage as "Lady Fairy".

Her program was inspired by the punch cards used in mechanical looms.

17

WORKED WITH THE POOR TO FIGHT SOCIAL INJUSTICE WITH MEDICINE.

FIRST WOMAN IN AMERICA TO RECEIVE A MEDICAL DEGREE.

FOUNDED THE NEW YORK INFIRMARY FOR WOMEN AND CHILDREN AND THE LONDON SCHOOL OF MEDICINE FOR WOMEN.

"IF THE PRESENT ARRANGEMENTS OF SOCIETY WILL NOT ADMIT OF WOMAN'S FREE DEVELOPMENT, THEN SOCIETY MUST BE REMODELLED." —ELIZABETH BLACKWELL

ELIZABETH BLACKWELL
DOCTOR

Elizabeth Blackwell had no interest in medicine until a friend of hers died from what was most likely uterine cancer. Her friend said she might have experienced less pain and suffering if only she had had a female doctor. This put Elizabeth on the path to becoming the first woman medical doctor in the United States.

Elizabeth was born into a family of abolitionists in 1821, with an upbringing that valued justice and equality. While working as a school teacher, she was mentored by male doctor friends and read books from their medical libraries. Although many didn't believe it was possible, she was accepted into Geneva Medical College.

Medical school is hard for any student, but Elizabeth faced additional challenges. Often met with hostility, she had to sit separately from the male students, and her teachers were embarrassed by her presence during anatomy lessons. When asked to leave a lecture about reproduction to protect her "delicate sensibilities," she argued her way into staying. During the summer she worked in a hospital in Philadelphia and saw how the hospital conditions contributed to the spread of infectious disease. The experience inspired her thesis on how good hygiene could prevent the spread of typhus. In 1849, she graduated from Geneva Medical College, first in her class.

Elizabeth's sister, Emily, also became a doctor. Together with Dr. Marie Zakrzewska, they opened the New York Infirmary for Indigent Women and Children in 1857. It was a place for the poor to get treatment and for female medical students and nurses to learn.

In the 1800s, there was little known about communicable diseases, and hand-washing was not mandatory for doctors like it is today. It was very common for doctors to go straight from treating someone with the flu to delivering a baby without even washing up. This caused the spread of diseases like typhus. Elizabeth realized that "prevention is better than cure," and in her lectures she advocated for better hygiene standards in hospitals and homes. Elizabeth went on to found the Woman's Medical College of the New York Infirmary in 1868 and the London School of Medicine for Women around 1874. An inspiration to many women, she also made it possible for many of them to become doctors.

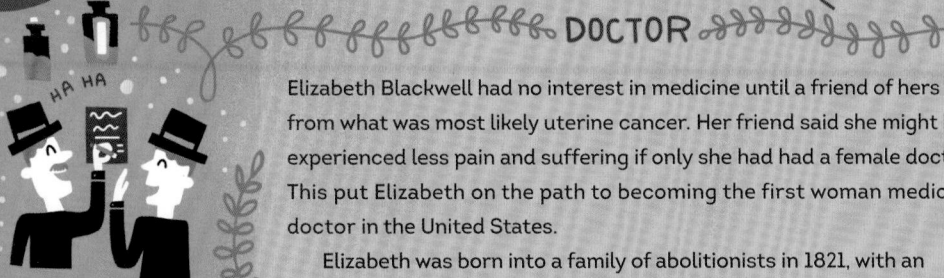

SHE WAS ACCEPTED INTO MEDICAL SCHOOL WHEN THE STUDENT BODY VOTED YES, AS A PRACTICAL JOKE; SHE SHOWED UP ANYWAY.

WAS A PROFESSOR OF GYNECOLOGY AT THE LONDON SCHOOL OF MEDICINE FOR WOMEN.

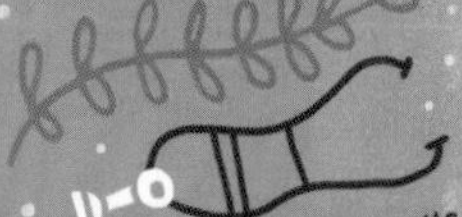

ADVOCATED FOR WOMEN'S RIGHTS, ESPECIALLY EQUAL OPPORTUNITY FOR FEMALE DOCTORS.

WROTE MANY BOOKS & PAPERS ON PUBERTY, PARENTING, AND FAMILY PLANNING.

TRAINED IN PARIS AND LONDON MATERNITY WARDS AFTER MEDICAL SCHOOL.

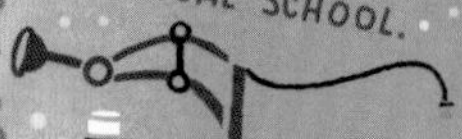

"I CAN NEVER BE A SURGEON NOW."

IN 1849, WHILE CARING FOR A BABY'S GONORRHEA-INFECTED EYE, ELIZABETH BECAME INFECTED HERSELF & LOST SIGHT IN ONE EYE.

HELPED TRAIN UNION NURSES WITH HER SISTER DURING THE CIVIL WAR.

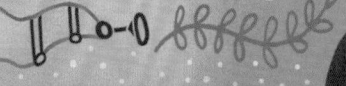

STARTED THE NATIONAL HEALTH SOCIETY IN LONDON.

"An error that ascribes to a man what was actually the work of a woman has more lives than a cat." — Hertha Ayrton

- First woman to win a Hughes Medal from the Royal Society in the UK.
- Invented a better electric arc and furthered our understanding of electrical current.
- First woman accepted into the Institution of Electrical Engineers.

HERTHA AYRTON

ENGINEER, MATHEMATICIAN, AND INVENTOR

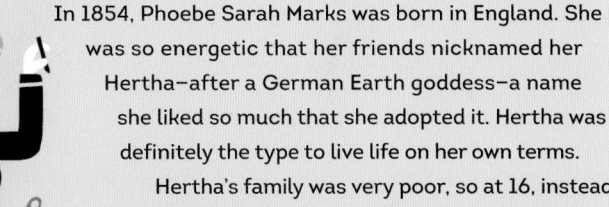

In 1854, Phoebe Sarah Marks was born in England. She was so energetic that her friends nicknamed her Hertha—after a German Earth goddess—a name she liked so much that she adopted it. Hertha was definitely the type to live life on her own terms.

Hertha's family was very poor, so at 16, instead of pursuing her passion to go to a university, she worked as governess to send money home. Fortunately, she met Madame Bodichon, a leader of the suffragist movement in the UK, who would help Hertha and pay for her education. In technical school, she met professor William Ayrton, who would become her husband and partner in invention.

In the 1890s, awful flickering and hissing electric arcs were used for streetlights and lighting in theaters. William and Hertha wanted to improve the lighting technology to make something quieter. At one point in the invention process, all of their notes accidentally burned in the fireplace, and Hertha had to start over from scratch. While William was away, she invented a new rod that made a clean and quiet bright light. Hertha burst open doors for women by getting published and giving lectures on electricity. During demonstrations about the arc, people were amazed to see a woman wielding such dangerous-looking equipment!

She was the first female member of the Institution of Electrical Engineers. However, women were not allowed to speak at the Royal Society. When her book *The Electric Arc* was published in 1902, it became too successful to ignore, and the Royal Society eventually allowed her to present her own paper. In 1906 they also awarded her the Hughes Medal for her body of work concerning electricity.

Hertha was also fearless when it came to politics. She was a vocal advocate of the suffragist movement and provided aid to female protesters on hunger strikes. Hertha participated in the 1911 boycott of England's census and wrote an impassioned letter on the form, demanding the vote for women!

Hertha's genius paved the way for women everywhere to play with "dangerous" machinery and invent great things!

REGISTERED 26 PATENTS.

WAS GOOD FRIENDS WITH MARIE CURIE.

STUDIED WIND MOTION AND WATER VORTICES.

NAMED HER CHILD AFTER MADAME BARBRA BODICHON, HER FRIEND & SUPPORTER.

WAS THE FIRST WOMAN NOMINATED TO BE A FELLOW OF THE ROYAL SOCIETY (THOUGH THEY DID NOT OFFICIALLY ACCEPT WOMEN UNTIL THE 1940s).

INVENTED THE AYRTON FAN TO BLOW AWAY MUSTARD GAS DURING WORLD WAR I.

INVENTED A LINE DIVIDER FOR ARCHITECTS.

CREATED A NEW THEORY OF NEUROSIS TO HELP PEOPLE COPE WITH ANXIETY.

HELPED CREATE A NEW FIELD OF PSYCHOLOGY CALLED NEO-FREUDIANISM.

DEVELOPED THE FOUNDATIONS OF FEMINIST PSYCHOLOGY.

"FORTUNATELY ANALYSIS IS NOT THE ONLY WAY TO RESOLVE INNER CONFLICTS. LIFE ITSELF STILL REMAINS A VERY EFFECTIVE THERAPIST." — KAREN HORNEY

KAREN HORNEY

PSYCHOANALYST

Karen Horney was born in Germany in 1885. In the early 1900s, psychology emerged as a new social science that researched how the mind worked. Sigmund Freud was the father of psychoanalytic theory, and his ideas were the basis of how everyone practiced at the time. Freudian theory focused mainly on male minds and posited that women wished they were men and therefore suffered from "penis envy."

Karen studied medicine at many schools, including The University of Berlin, where she earned her medical degree. Her own battles with depression inspired her to study psychology. She was mentored and analyzed by Karl Abraham and was well versed in Freudian theory. She started treating her own patients and began officially teaching at the Berlin Psychoanalytic Institute in 1920. Through her many clinical studies, she began to observe behavior that did not fit the framework of Freudian theory, leading her to rebel against everything she had been taught.

Karen argued that society didn't allow for women to have any real power but instead forced them to live through their husbands and children. She theorized that women didn't want to become men; they just wanted the independence that men had. She argued that society shapes a person's perception of self-worth. In doing this, she created the field of feminist psychology.

Karen moved to America in 1932 and worked with the New School for Social Research and the New York Psychoanalytic Institute. There, she created a new theory of neurosis. She realized that anxiety is not just shaped by our biological urges but also is caused by the environment in which we grow up. This Neo-Freudian therapy meant that people could learn to cope with their anxieties and eventually not need therapy. This directly contradicted Freud's theories, and Karen faced a fierce backlash, which eventually forced her out of the New York Psychoanalytic Institute in 1941. Despite this, she continued to write many books and papers and founded the Association for the Advancement of Psychoanalysis.

Karen Horney created a new way of thinking about ourselves, society, and anxiety. She is still considered one of the most influential psychologists ever.

Her mentor, Karl Abraham, was a super close friend of Freud's.

Founded the American Journal of Psychoanalysis.

Founded the American Institute for Psychoanalysis & became the dean.

She wrote many books including the popular The Neurotic Personality of Our Time.

Inspired the term "womb envy."

The Horney Clinic in New York is named after her.

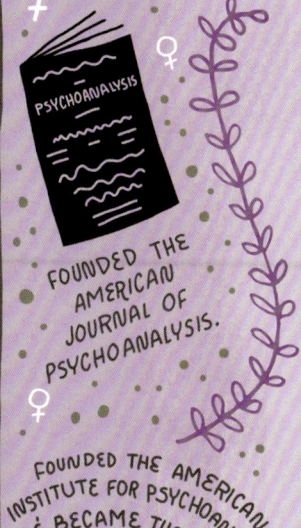

DISCOVERED THAT SEX IS DETERMINED BY "X" AND "Y" CHROMOSOMES.

ONE OF THE FIRST WOMEN IN AMERICA TO BE RECOGNIZED FOR HER BIOLOGY RESEARCH.

CHANGED HOW WE STUDY EMBRYOS AND CYTOGENETICS.

"MISS STEVENS HAD A SHARE IN A DISCOVERY OF IMPORTANCE, AND HER WORK WILL BE REMEMBERED FOR THIS." —THOMAS HUNT MORGAN

NETTIE STEVENS

GENETICIST

Nettie Stevens was born in 1861 in Vermont. She pinched pennies to pay for her education, and often taught classes to help pay her way. She had a very long road to follow to get to her short but groundbreaking career. Nettie slowly finished her undergrad education at the newly formed Stanford University in California. After earning her master's degree, her interest in genetics brought her back to the East Coast, where she received her PhD at Bryn Mawr College at the age of 41.

The big question in genetics at the time was a simple one: what makes a baby a girl or a boy? At the time, sex determination was a still a mystery. For centuries doctors thought sex was determined by what a woman ate during pregnancy or how warm she kept her body. Nettie and other scientists had their suspicions that there was more to sex determination than that.

Nettie got to work by dissecting bugs. She took the sex organs from butterflies and mealworms to look at the cells under a microscope. Male insects had an XY-shaped chromosome, and females had an XX. Her flawless technique and use of different kinds of bugs strengthened the hypothesis she made based on her observations. In 1905 she published her groundbreaking research in a 2-part book, which overturned hundreds of years of misconceptions.

Around the same time, Edmund Wilson, Nettie's former advisor, made the same discovery of XY chromosomes on his own, but Nettie's work had the strongest proof. She wrote about her findings with great scientific conviction, but it was received by a skeptical public. Unfortunately, her untimely death in 1912 has rendered her largely overlooked and forgotten.

We now recognize Nettie for her amazing work, which allowed scientists to better understand sex determination and genetics.

She also used fruit flies & beetles in her studies.

Her dad was a carpenter.

Her historic work was called Studies in Spermatogenesis.

Traveled to Italy and Germany to study cytology.

To make sure their baby would be a boy, people used to try to conceive in the summer (it didn't work).

Nobel Prize winner Thomas Morgan's work was possible because of Nettie's research.

FIRST WOMAN TO WORK FOR THE US GEOLOGICAL SURVEY.

TRAINED ALMOST EVERY FEMALE GEOLOGIST OF HER TIME.

WAS AN EXPERT ON THE PIEDMONT PLATEAU.

"I HAVE CONSIDERABLE PRIDE IN THE FACT THAT SOME OF THE BEST WORK DONE IN GEOLOGY TODAY BY WOMEN, RANKING WITH THAT DONE BY MEN, HAS BEEN DONE BY MY STUDENTS." —FLORENCE BASCOM

FLORENCE BASCOM

GEOLOGIST AND EDUCATOR

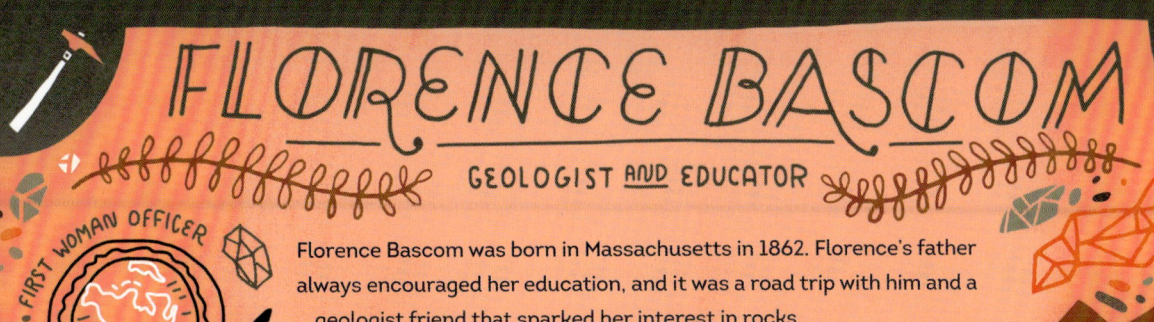

Florence Bascom was born in Massachusetts in 1862. Florence's father always encouraged her education, and it was a road trip with him and a geologist friend that sparked her interest in rocks.

In 1893, Florence was the first woman to get a PhD from Johns Hopkins University, and it did not come easily. She was forced to take her classes behind a screen so she wouldn't "distract" any of her male classmates. Despite the unfair treatment she received, she loved learning and became the second woman in America to complete a geology doctorate. She would go on to inspire many more female geologists.

Florence became an authority on rocks and how to categorize them through their chemical makeup and mineral content. By studying the layers in rocks, we gain a better understanding of the history and evolution of our planet's surface. In her dissertation, Florence's expertise allowed her to prove that a layer of rock everyone had thought was sedimentary was actually caused by lava flows.

Florence began teaching and researching at Bryn Mawr College in 1895, where she founded the geology program, and her curriculum was one of the most highly regarded in the nation. She trained almost every female geologist in America until she retired in 1928. Working at a women's college gave Florence research opportunities she might not have otherwise had. She was a rigorous teacher but also was able to do important field work for the US Geological Survey. At Bryn Mawr, she began her intensive work in geomorphology, the study of how the earth's geography changes over thousands and thousands of years. Florence's research focused on the hilly area of the Appalachians, also known as the Mid-Atlantic Piedmont. She created important geographical maps of New Jersey and Pennsylvania that are still used today.

Florence Bascom rocked the world of geology! Her discoveries and maps continue to influence the field.

First woman officer of the Geologic Society of America.

Associate editor of The American Geologist.

Worked in a storage space where she collected fossils & rocks.

Received 4 stars in the first edition of American Men of Science in 1906.

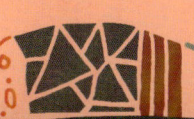

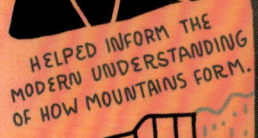

She studied Philadelphia's water resources.

Published over 40 scientific papers.

Helped inform the modern understanding of how mountains form.

PIONEERED RESEARCH ON RADIOACTIVITY.

WON TWO NOBEL PRIZES.

FOUNDED THE CURIE INSTITUTE IN PARIS.

DISCOVERED 2 ELEMENTS: POLONIUM AND RADIUM.

"I WAS TAUGHT THAT THE WAY OF PROGRESS IS NEITHER SWIFT NOR EASY." — MARIE CURIE

MARIE CURIE
PHYSICIST AND CHEMIST

Marie Curie was born in Warsaw, Poland, in 1867. After working as a governess to support her sister's education, it was Marie's turn. She traveled to Paris to study at the Sorbonne, where she met Pierre Curie, a fellow scientist and her greatest love.

Scientist Henri Becquerel had discovered a mysterious glow coming from uranium salts. Scientists didn't seem too interested in the effect, but Marie was fascinated by the glow and wanted to know what it was and why it was happening. In a stuffy shed, Marie and Pierre went to work. Using Pierre's electrometer, Marie examined "glowing" compounds and discovered that the energy being produced came from the uranium atom itself! We now know that atoms with an unstable nucleus emit particles and release energy. Marie started calling the effect "radioactivity." To find the source, she and Pierre ground up and filtered down other radioactive materials, like the mineral ore uraninite. Through this process Pierre and Marie discovered 2 new radioactive elements: polonium and radium. Together, the Curies received a Nobel prize in physics in 1903 for the discovery of radiation. Later, in 1911, Marie won a second Nobel prize in chemistry for her discovery of and research into polonium and radium.

Pierre and Marie made an amazing team. Sadly, they realized that the radiation from their experiments was making them sick. Pierre would do tests with radium on his own arm that left large burns. Their long-term exposure made them both tired and achy—we now understand that the effects of radiation poisoning are deadly. In 1906, Pierre was killed in a horse-carriage accident. Despite her grief and the danger involved, Marie continued their important work and discovered that radium could be used as a cancer treatment. She spent hours collecting radon gas to send to hospitals even though it left her feeling weak.

In 1914, France was invaded during World War I. With her daughter Irène, she created a unit of mobile medical X-ray trucks, which they heroically drove onto the battlefields to help wounded soldiers.

Marie Curie did scientific work because she loved it and dangerous work because the world needed it. Her life and work continue to inspire scientists today.

- FIRST WOMAN TO GET A DOCTORATE IN FRANCE.
- POLONIUM WAS NAMED AFTER POLAND.
- RADIUM WAS NAMED AFTER THE SUN.
- MOTHER OF 2 GIRLS.
- FIRST WOMAN TO BE HONORED FOR HER OWN ACHIEVEMENTS WITH HER BURIAL IN THE PANTHEON IN PARIS.
- ONLY PERSON TO WIN A NOBEL IN TWO DIFFERENT DISCIPLINES.
- COINED THE WORD "RADIOACTIVITY."
- ALL OF HER RESEARCH IS KEPT IN LEAD-LINED CASES. THE MATERIALS ARE STILL RADIOACTIVE.
- KEPT VIALS OF GLOWING RADIUM IN HER POCKETS, A DANGEROUS PRACTICE.
- INHERITED PIERRE'S CHAIR AT THE SORBONNE, BECOMING THEIR FIRST FEMALE PROFESSOR.

SUFFRAGIST WHO FOUGHT FOR WOMEN'S RIGHT TO VOTE.

WORLD'S GREATEST AGROSTOLOGIST (EXPERT IN GRASS).

IDENTIFIED THOUSANDS OF TYPES OF GRASS ALL OVER THE WORLD.

"GRASS MADE IT POSSIBLE FOR THE HUMAN RACE TO ABANDON CAVE LIFE AND FOLLOW HERDS." — MARY AGNES CHASE

MARY AGNES CHASE

BOTANIST AND SUFFRAGIST

Mary Agnes Chase was a tiny woman with a fighting spirit. She was born in 1869 and grew up in Chicago. She started working after finishing grammar school in order to help her family, but in her spare time, she enjoyed learning about botany. She would go on trips to sketch plants and used her small savings to take a few botany classes at the University of Chicago and the Lewis Institute. Her informal education also included working with botanist Reverend Ellsworth Jerome Hill; he mentored Mary, and in exchange she illustrated plants for his papers.

Her impressive sketchbooks got her a part-time job at the Chicago Field Museum of Natural History, where she was the scientific illustrator for a few of the museum's publications. Mary figured out how to use a microscope and do technical drawings on the job. With her new skills, Mary became a full-time illustrator for the United States Department of Agriculture (USDA) in 1903.

At the USDA Mary worked as assistant to the botanist Albert Hitchcock. Together they took on the task of collecting and classifying grasses in North and South America until his death in 1935, when she became the senior botanist in charge of systematic agrostology. Unlike her male colleges, Mary was often denied funding to travel, but, not content to just stay in the lab, she traveled all over the United States and South America, even if it meant paying her own way. Mary discovered thousands of new species of grasses from around the world and authored and coauthored many books on those plants.

Mary called grass "the plant that holds the soil," and she was able to figure out which grasses were the best to feed livestock. With Albert Hitchcock, she studied commercially developed grass strains to make sure that they were as advertised. A lot of today's food has been informed by Mary's important research.

Mary was also a suffragist. She protested for women's right to vote in the United States even when the USDA threatened to fire her. She bravely participated in the 1918 hunger strike, in which she was jailed and force-fed. Her sacrifices helped gain women the right to vote in 1920.

Mary continued to work for the USDA until she retired in 1939. She was an honorary curator for the Smithsonian up until her death in 1963. Her research was left to the Smithsonian, where it continues to be used.

WORKED ODD JOBS IN STOCKYARDS, A GROCERY STORE, AND A MAGAZINE.

SOLO WROTE & ILLUSTRATED A FIRST BOOK OF GRASSES: THE STRUCTURE OF GRASSES EXPLAINED FOR BEGINNERS.

GIVEN AN HONORARY DEGREE FROM THE UNIVERSITY OF ILLINOIS.

WAS AN ACTIVE MEMBER IN THE NAACP.

WAS AN HONORARY FELLOW AT THE SMITHSONIAN INSTITUTION AND FELLOW AT THE LINNEAN SOCIETY OF LONDON.

HER HOME IN WASHINGTON, D.C. CALLED CASA CONTENTA BECAME A PLACE FOR LATIN AMERICAN WOMEN BOTANISTS TO STAY WHILE LEARNING IN THE U.S.

COLLECTED OVER 10,000 DIFFERENT TYPES OF GRASS SPECIMENS FROM AROUND THE WORLD.

TIMELINE

Throughout history, many obstacles have stood in the way of women who pursued the sciences. A lack of access to higher education and not being paid a fair wage are just some of those barriers. Let's celebrate the milestones in history and accomplishments women have made in education and science.

1780s
Caroline Herschel, astronomer, was the first woman to become an honorary member of the Royal Society.

1833
Oberlin College was the first college in America to admit women.

1903
Marie Curie was the first woman to receive a Nobel prize.

1947
Marie Daly became the first African-American woman to earn a PhD in chemistry.

1955-72
The Space Race between the United States and the USSR caused a boom of innovation and engineering opportunities for women and men.

1963
Valentina Tereshkova was the first woman in space.

400 AD
Hypatia of Alexandria was the first recorded female mathematician.

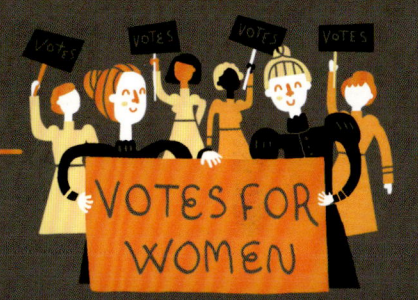

1678
Elena Piscopia was the first woman in the world to receive a doctoral degree.

1715
Sybilla Masters was the first woman in the United States to get a patent for her invention, which cleaned and processed corn.

1920
Women gained the right to vote in the United States with the Nineteenth Amendment.

1941-45
WWII created a new workforce of women while men were at war. Female scientists were given new opportunities to show their talents.

1946
An all-female team programed the first all-electronic computer with the Electronic Numerical Integrator And Computer (ENIAC) project.

1963
The Equal Pay Act passed in America and stipulated that men and women should be paid equally for equal work. This law helps women to overcome the wage gap (the fight is ongoing).

1964
The Civil Rights Act made many forms of discrimination illegal, ending racial segregation in schools and workplaces and giving more opportunities to African-Americans.

NOW
More women than ever before are working hard to invent, discover, and explore the unknown.

DISCOVERED AND EXPLAINED THE WORKINGS OF NUCLEAR FISSION.

DISCOVERED THE ELEMENT PROTACTINIUM WITH HER LAB PARTNER OTTO HAHN.

SHOULD HAVE RECEIVED A NOBEL PRIZE.

"LIFE NEED NOT BE EASY, PROVIDED ONLY IT WAS NOT EMPTY." —LISE MEITNER

LISE MEITNER
PHYSICIST

Lise Meitner was born in 1878. Like many Jewish families at the time, hers was living happily in Vienna. Lise loved science but knew that as a girl, she would need to fight to be able to pursue an education.

After Lise received her PhD, she went to work at the chemistry institute in Berlin in 1907. There, she met Otto Hahn, who would become her collaborator throughout her career. Even though she was brilliant, being a woman meant she was unpaid and not allowed to use the labs or even the bathrooms. Until the government officially permitted women to attend the university, she did all of her radiochemistry research in a dank basement.

In 1934, scientists focused on discovering new heavy elements. Lise and Otto were trying to artificially create new elements by smashing neutrons against uranium. They did not know it yet, but they were on the brink of a brand-new discovery. Lise's research was interrupted by the Nazis' rise to power. Because she was Jewish, Lise needed to escape but did not want to leave her work. In 1938, with a heavy heart, she fled to Sweden and Otto continued their work in Germany.

She and Otto secretly wrote letters back and forth about their research. He struggled to understand the results of their experiments. Lise realized that they were not creating a new element, but that their work was causing the nucleus of one atom to stretch apart and release energy. From afar, Lise discovered nuclear fission—the nuclear reaction that releases nuclear energy.

Lise was unable to return to Germany, and Otto was awarded the 1944 Nobel Prize for their work—without her. Lise refused to ever work in Germany again, unable to forgive the country for what it had done to her people.

Although she did not win the Nobel prize, Lise wrote papers on fission that were read all over the world, and she won many other awards. Her brilliant mind gave us a new form of energy and changed physics forever.

She's "the German Marie Curie!"

SHE KNEW ALBERT EINSTEIN.

HELPED AUSTRIA IN WWI AS AN X-RAY NURSE.

ENERGY FROM THE FISSION EXPERIMENT WAS DESCRIBED AS 20 MILLION TIMES MORE POWERFUL THAN TNT.

109 Mt

HAD AN ELEMENT, MEITNERIUM, NAMED IN HER HONOR.

COMPARED NUCLEAR FISSION TO THE STRETCHING OF PIZZA DOUGH.

ESCAPED FROM GERMANY WITH THE HELP OF PHYSICIST NIELS BOHR.

DINED WITH PRESIDENT TRUMAN AS THE WOMAN OF THE YEAR.

PIONEER IN ERGONOMICS, TIME AND MOTION STUDIES, AND ORGANIZATIONAL PSYCHOLOGY.

FIRST WOMAN IN THE AMERICAN SOCIETY OF MECHANICAL ENGINEERS.

REINVENTED THE MODERN KITCHEN SPACE.

"WE CONSIDERED OUR TIME TOO VALUABLE TO BE DEVOTED TO ACTUAL LABOR IN THE HOME. WE WERE EXECUTIVES." —LILLIAN GILBRETH TALKING TO A GROUP OF BUSINESS WOMEN

LILLIAN GILBRETH
PSYCHOLOGIST AND INDUSTRIAL ENGINEER

Lillian Gilbreth was born in 1878 into a big family of 9 children. She was always interested in her education and graduated from the University of California, Berkeley, with a master's in literature.

She met Frank Gilbreth in the middle of completing her PhD. She was intrigued by his obsession with workplace efficiency. She switched her major from literature to psychology and wrote her dissertation, "The Psychology of Management." It was the first study of organizational psychology and how relationships affect us at work. She earned her PhD from Brown University in 1915.

Together Lillian and Frank ran a consulting business. They would study a simple task, like bricklaying or carrying tools, and break the motions down to the most essential steps to make the workers' jobs easier and quicker.

Lillian authored and coauthored (with Frank) many books about motion and fatigue. Often, only Frank's name would appear on their work because publishers thought a male author would appear more authoritative and credible—even though she was the educated psychologist.

When Frank died in 1924, Lillian took sole charge of their company. Many of their clients did not want a woman telling them how to run their factories. Since they thought that women belonged in the kitchen, that is what Lillian decided to focus on: homemakers. Back then, it was common for women to spend all day long cooking and cleaning. It was backbreaking and exhausting labor. Lillian wanted to apply ergonomics and motion studies to help make housewives' domestic jobs easier. She created new tools and a new layout for kitchens that cut work time down from an entire day to only a few hours. This gave women all over the country more time to pursue more stimulating interests.

Lillian continued to be the president of her firm, working with all sorts of clients. She even helped the United States government create jobs during the Great Depression with the President's Organization for Unemployment Relief.

Look around you and you are likely to see something Lillian Gilbreth designed to save you time. Whether it is the ergonomic layout of your desk or the "work triangle" that determines the distance from the sink to the stove, Lillian's Gilbreth's designs have been integrated into our daily lives.

TESTED OUT NEW EFFICIENCY TECHNIQUES ON HER 12 CHILDREN.

CALLED MOVEMENT UNITS "THERBLIGS" (GILBRETH SPELLED BACKWARDS).

INVENTED THE FOOT PEDAL ON THE GARBAGE CAN AND SHELVES IN THE FRIDGE.

TESTED HER NEW KITCHEN SYSTEM BY MAKING STRAWBERRY SHORTCAKE.

USED HER KNOWLEDGE OF ERGONOMICS TO HELP HANDICAPPED MEN AND WOMEN FIND WORK.

RECEIVED MANY HONORARY DEGREES.

NICKNAMED THE "FIRST LADY OF MANAGEMENT".

CREATED THE FIELD OF ABSTRACT ALGEBRA.

$$j = \sum_{i=1}^{3} \frac{\partial L}{\partial \dot{x}_i} Q[x_i] - f$$
$$= m\sum_i \dot{x}_i^2 - \left[\frac{m}{2}\sum_i \dot{x}_i^2 - V(x)\right]$$
$$= \frac{m}{2}\sum_i \dot{x}_i^2 + V(x).$$

THE NOETHER THEORY CONNECTS MATHEMATICAL SYMMETRY TO THE CONSERVATION OF ENERGY.

CONSIDERED ONE OF THE MOST IMPORTANT PEOPLE IN THE FIELD OF MATHEMATICS.

$$\left(\frac{\partial L}{\partial \dot{q}}\dot{q} - L\right)T - \frac{\partial L}{\partial \dot{q}}\frac{\partial \phi}{\partial \varepsilon}$$

"MY METHODS ARE REALLY METHODS OF WORKING AND THINKING; THIS IS WHY THEY HAVE CREPT IN EVERYWHERE ANONYMOUSLY." —EMMY NOETHER

EMMY NOETHER
MATHEMATICIAN AND THEORETICAL PHYSICIST

Emmy Noether was born in Germany in 1882. She grew up in a family of mathematicians and wanted to learn like her father and brothers. At the time, it was against the law in Germany for women to get a higher education, so she would sit in the back of classes at the university to try to learn as much as she could while receiving no academic credit. For over 2 years she audited classes until they finally admitted her as a student. At the University of Erlangen, Emmy lectured unofficially in her father's classrooms and worked without pay or title. She started to make waves in the physics community with the half-dozen papers she published and her talks abroad. Around 1915, Albert Einstein's team recruited her to the University of Göttingen to help further develop his general theory of relativity. He became a friend who would always defend Emmy.

Emmy worked for free for 7 years at Göttingen until she finally started getting paid, but she was the lowest-paid professor. Despite the lack of recognition, she developed mathematical equations that are still an important part of the way we understand physics now. She produced developments in the field of abstract algebra by proving new concepts about groups and rings. She made new connections between energy and time, and angular momentum. In doing all of this, she developed the Noether theory.

Because Emmy was Jewish, the rise of the Nazi regime put her life in danger. She was fired from Göttingen for being Jewish but continued to teach from her apartment in secret. In 1933, Emmy escaped to America, where she was hired to teach at Bryn Mawr College. Unfortunately, only 18 months after she began teaching with good pay and a real title, she became ill and died at the age of 53.

After her death, Albert Einstein made sure she would be remembered. In 1935, he wrote to the *New York Times* that "Fraulein Noether was the most significant mathematical genius thus far produced since the higher education of women began."

- HER STUDENTS WERE CALLED "NOETHER BOYS."
- "IF I DON'T EAT I CAN'T DO MATHEMATICS!"
- PEOPLE WOULD MAKE FUN OF HER WEIGHT AND APPEARANCE.
- WAS A PACIFIST, DESPITE THE PERSECUTION SHE FACED IN WWII.
- HER FATHER, MAX NOETHER, WAS ALSO AN IMPORTANT MATHEMATICIAN.
- SCHOOLS AND A MOON CRATER ARE NAMED AFTER HER.
- HER ASHES WERE BURIED AT BRYN MAWR.

- Created some of the first "software" for electrical engineering.
- Invented a graphical calculator to help solve equations involving hyperbolic functions.
- Expert in equivalent circuits and graphical analysis.
- First female electrical engineer.

"THERE IS NO DEMAND FOR WOMEN ENGINEERS, AS SUCH, AS THERE ARE FOR WOMEN DOCTORS; BUT THERE'S ALWAYS A DEMAND FOR ANYONE WHO CAN DO A GOOD PIECE OF WORK." —EDITH CLARKE

EDITH CLARKE
ELECTRICAL ENGINEER

Edith Clarke was born in Maryland in 1883. Tragedy struck when both of her parents died before she turned 12. She used the money she inherited to pay for college and would never let anything hold her back from becoming an electrical engineer.

After earning her bachelor's degree from Vassar, Edith spent some time studying at the University of Wisconsin-Madison. She paused her studies to start work as a human computer for AT&T. Before mechanical computers, engineers and scientists would rely on a group of people crunching complicated math formulas to aid them in their work. At the time, human "computing" was seen as women's work and engineering was seen as men's work.

Determined to finish her education, Edith left her job and enrolled at the Massachusetts Institute of Technology (MIT). In 1919, she became the first woman to graduate from MIT with a master's degree in electrical engineering. Still she could only find work crunching numbers.

General Electric hired her to calculate and train other women. While working as a human calculator, she invented a new graphical calculator, and because she was only a part-time employee, GE missed out on claiming the rights to it. She filed a patent in 1921, which became official in 1925. Now equations with hyperbolic functions could easily be solved.

GE still wouldn't recognize her as an engineer, so the same year she invented her calculator, she quit. For a year she taught in Constantinople (now Istanbul), Turkey, and traveled the world. Her absence must have made an impression, because when she returned in 1922, GE hired her as the first official female electrical engineer.

Edith continued to create more efficient methods of calculating equations. She made it easier for engineers to manage large, complicated power systems. She also figured out how to get the most power possible out of transmission lines.

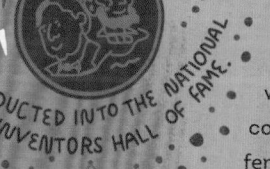

Edith retired from GE in 1945 and went on to teach at the University of Texas for 10 years. Her work gained respect in the electrical engineering community, and in 1948 she became the first female Fellow of the American Institute of Electrical Engineers (AIEE). Edith Clarke was a trailblazer and proved that a woman can definitely do "a man's job."

- FIRST WOMAN TO BE ALLOWED TO SHARE HER PAPER WITH THE AMERICAN INSTITUTE OF ELECTRICAL ENGINEERS.
- PUBLISHED 18 TECHNICAL PAPERS IN 22 YEARS.
- WON THE SOCIETY OF WOMEN ENGINEERS ACHIEVEMENT AWARD IN 1954.
- INDUCTED INTO THE NATIONAL INVENTORS HALL OF FAME.
- GREW UP WITH A READING AND WRITING LEARNING DISABILITY.
- WROTE ONE OF THE MOST IMPORTANT BOOKS ON ELECTRICAL ENGINEERING: CIRCUIT ANALYSIS OF A-C POWER SYSTEMS.
- FIRST FEMALE PROFESSOR IN HER FIELD IN THE US.
- HELPED DESIGN HYDROELECTRIC DAMS.

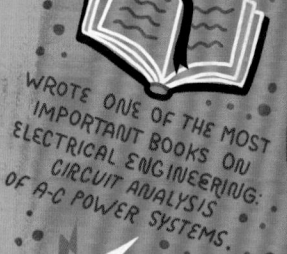

"I'D LIKE TO HEAR LESS TALK ABOUT MEN AND WOMEN AND MORE TALK ABOUT CITIZENS." —MARJORY STONEMAN DOUGLAS

GAVE US NEW INSIGHTS INTO THE ECOSYSTEMS OF WETLANDS.

CONSERVATIONIST, SUFFRAGIST, AND ADVOCATE FOR CIVIL RIGHTS.

FOUNDED THE FRIENDS OF THE EVERGLADES.

HER WORK HELPED TO ESTABLISH EVERGLADES NATIONAL PARK.

MARJORY STONEMAN DOUGLAS

WRITER AND CONSERVATIONIST

In the late 1940s, Florida's Everglades were seen as a nuisance, just one big swamp that needed draining. The only thing that prevented the destruction of the wetlands was a feisty woman named Marjory Stoneman Douglas.

Marjory was born in 1890 in Minneapolis and graduated from Wellesley College. She always wanted to be a writer and, after ending a bad marriage, she got a job at the *Miami Herald* where her father also worked. She started her career as a society reporter in 1915.

Her father used his status as editor of the newspaper to talk about politics and criticize the governor's plan to drain the Everglades. As a result, Marjory understood how powerful words could be, and she began to use her own writing to talk about civil rights, the suffrage movement, and environmental conservation.

Ernest Coe, a fellow conservationist, asked Marjory to help save the Everglades. Although the land was no place for a picnic, "too buggy, too wet," Marjory fell in love with its natural beauty. She discovered that the Everglades was not just a swamp, but a river that is vital to Florida's ecosystem. She published *The Everglades: River of Grass* in 1947. The book and Marjory became famous. Her work led directly to the creation of Everglades National Park.

Although the government began to protect the Everglades, Marjory needed to protect the land from the US Army Corps of Engineers, whose agricultural dams and canals were disrupting the ecosystem. A proposed jetport project threatened its destruction. Marjory's gumption and expert knowledge of the land ensured her a win. In 1969, she started the "Friends of the Everglades" organization and halted construction.

Marjory continued her work well into the 1990s. Despite being nearly blind, she continued to write and fight for the Everglades. Her energy and passion only increased, and she was awarded the Presidential Medal of Freedom in 1993. She died at the age of 108 in 1998.

The Everglades is home to alligators, manatees, and many species of birds and fish.

Marjory understood that "there is no other Everglades in the world"

It is a unique and delicate ecosystem.

Her ashes were scattered over her national park.

The Everglades' wide, shallow waterway moves very slowly — a phenomenon called sheetflow.

Became known for her floppy hat and dark round glasses.

Worked as a Red Cross nurse in Europe during World War I.

INVENTED THE BALL METHOD.

FIRST AFRICAN-AMERICAN AND FIRST WOMAN TO GRADUATE FROM THE UNIVERSITY OF HAWAII.

HELPED TO CURE LEPROSY WITH HER CHEMICAL TREATMENT.

"MEN DOMINATED HIGHER EDUCATION IN 1915, AND ALICE BALL WAS ADMITTED AGAINST THE ODDS." – MILES JACKSON, UNIVERSITY OF HAWAII PROFESSOR AND DEAN EMERITUS

ALICE BALL
CHEMIST

Alice Ball was born in Seattle in 1892. Her grandfather was a famous photographer, and in his darkroom, Alice was introduced to the wonders of chemistry.

She began her formal education in chemistry at the University of Washington and then moved to Hawaii to earn her master's. In 1915 she became the first African-American and the first woman to graduate from the University of Hawaii.

In the early 1900s, there was a public health emergency. Leprosy, now known as Hansen's disease, was spreading—it causes numbness, skin lesions leading to permanent deformities, and damage to the nerves and eyes. To this day we aren't entirely sure how the disease spreads but we now know that it is not very contagious. Back then, police arrested the sick and isolated them in the Kalaupapa leper colony on the Hawaiian island of Molokai.

At the time, there was only one source of relief for leprosy: the thick, sticky oil of the chaulmoogra tree's seeds. But it was impossible to mix the oil with water in order to make a suitable treatment that could be injected—our blood is mostly water—and the oil by itself was ineffective and painful to inject. Rubbing it on the skin or swallowing it didn't work much better. Alice was on the case to figure out how to create an injectable cure.

At age 23, Alice developed a new way to treat the dense chaulmoogra oil. After isolating the ethyl esters in its fatty acids, she found the oil could be blended with water for injection. This new treatment, which became known as the "Ball method," helped the colony of people suffering from leprosy. No longer feared to be contagious, the sick did not need to be isolated. By 1918 patients could see their families, and new patients were no longer forced into exile.

Alice died, too soon and too young, in 1916 while teaching a lab. Many think she accidentally inhaled chlorine gas. She is now remembered for finding a cure for what seemed like a hopeless disease.

HER DAD WAS A FAMOUS LAWYER.

STARTING IN 1866 & INTO THE 20TH CENTURY, OVER 8000 PEOPLE WITH LEPROSY WERE SENT TO KALAUPAPA.

THE UNIVERSITY OF HAWAII HONORED ALICE WITH A PLAQUE ON A CHAULMOOGRA TREE.

CO-PUBLISHED A PAPER IN JOURNAL OF THE AMERICAN CHEMICAL SOCIETY WHILE IN COLLEGE.

CHAULMOOGRA OIL CAUSED MAJOR STOMACH PAIN WHEN SWALLOWED.

FEBRUARY 29TH, EVERY FOUR YEARS, IS ALICE BALL DAY IN HAWAII.

DEVELOPED THE ONLY WORKING TREATMENT FOR LEPROSY UNTIL ANTIBIOTICS WERE DEVELOPED IN THE 1940s.

"As a researcher the unforgotten moments of my life are rare ones... when the veil over nature's secrets seems to suddenly lift..." —Gerty Cori

- Her work has given us an understanding of carbohydrate metabolism.
- Codiscovered the Cori cycle.
- Won a Nobel Prize in Physiology or Medicine.

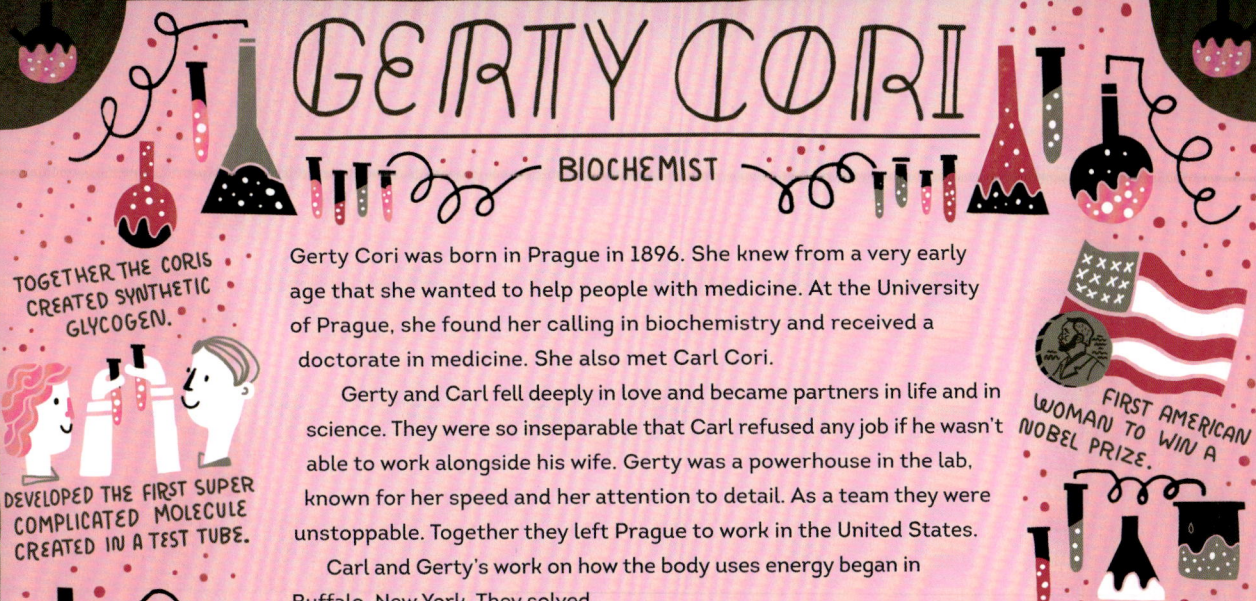

GERTY CORI
BIOCHEMIST

Gerty Cori was born in Prague in 1896. She knew from a very early age that she wanted to help people with medicine. At the University of Prague, she found her calling in biochemistry and received a doctorate in medicine. She also met Carl Cori.

Gerty and Carl fell deeply in love and became partners in life and in science. They were so inseparable that Carl refused any job if he wasn't able to work alongside his wife. Gerty was a powerhouse in the lab, known for her speed and her attention to detail. As a team they were unstoppable. Together they left Prague to work in the United States.

Carl and Gerty's work on how the body uses energy began in Buffalo, New York. They solved the mystery of how cells use sugar for energy. They figured out how our bodies convert glucose into lactate (and vice versa), using our muscles and liver. This allows us to use energy when we exercise and store energy for later.

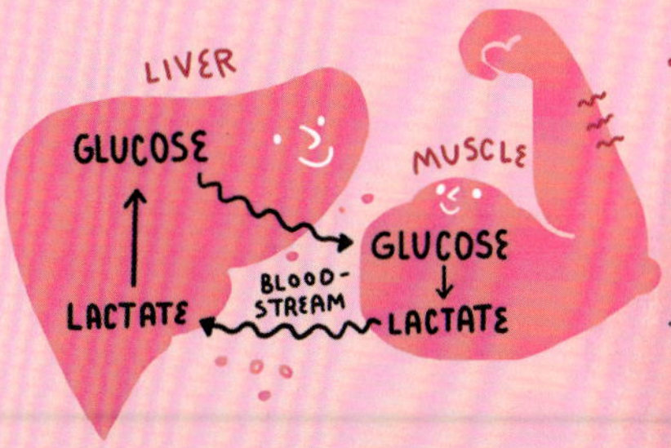

This process is called the Cori cycle, named for Gerty and Carl. They continued their work in their own laboratory at Washington University School of Medicine, which became a hot spot for biochemistry.

In 1947, Gerty and Carl shared a Nobel Prize for their amazing contributions to medicine. Soon after, Gerty developed a bone marrow disease but continued to work in the lab as always. When she became too weak to get around the laboratory, Carl would carry her where she needed to go. The only thing more important to them than their work was each other. Gerty died in 1957 at age 61.

TOGETHER THE CORIS CREATED SYNTHETIC GLYCOGEN.

DEVELOPED THE FIRST SUPER COMPLICATED MOLECULE CREATED IN A TEST TUBE.

TOGETHER THE CORIS PUBLISHED 50 PAPERS IN 9 YEARS.

STUDIED ENZYMES AND HORMONES RELATED TO PROCESSING SUGAR.

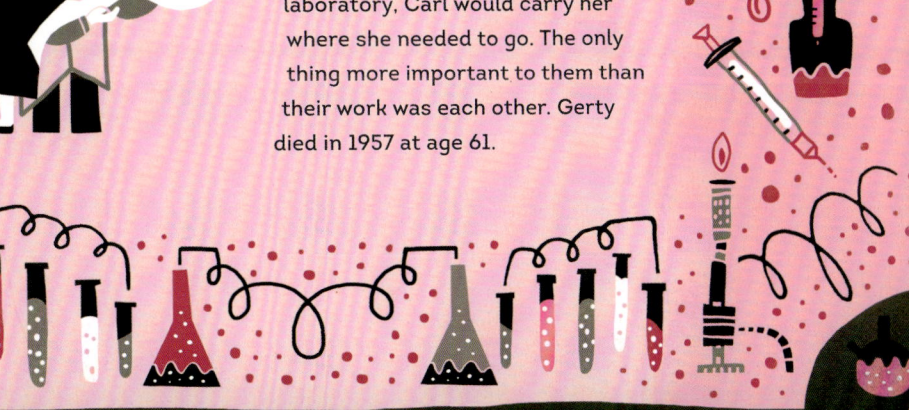

FIRST AMERICAN WOMAN TO WIN A NOBEL PRIZE.

THE CORIS' LABORATORY WAS THE TRAINING GROUND FOR 6 OTHER NOBEL PRIZE WINNERS.

HELPED US TO UNDERSTAND DIABETES.

LIVER: GLUCOSE ↑↓ LACTATE — BLOODSTREAM — MUSCLE: GLUCOSE ↑↓ LACTATE

DISCOVERED A NEW SPECIES, THE PENINSULA DRAGON LIZARD.

CONSIDERED AN EXPERT IN HERPETOLOGY.

DESIGNED THE MOST COMPLICATED AND ADVANCED REPTILE HOUSE OF ITS TIME.

"WHY SHOULDN'T A WOMAN RUN A REPTILE HOUSE? WOMEN ARE AT WORK IN MY COUNTRY, AND THE REST OF THE WORLD, IN ALL TYPES OF WORK AND PROFESSIONS." —JOAN PROCTER

JOAN BEAUCHAMP PROCTER
ZOOLOGIST

Joan Beauchamp Procter always had a fascination with reptiles. She was born in England in 1897 and grew up in a time when women were seen as dainty, and reptiles were considered exotic and dangerous. Joan's chronic ill health kept her from going to a university, but it didn't keep her from studying the animals she loved.

Joan kept snakes, frogs, and even a crocodile as pets. She used her animals to present a paper to the Zoological Society of London when she was only 19. In 1917, she started officially working at the British Museum as an assistant to George Albert Boulenger, keeper of the reptile and fishes. In 1923, she became the London Zoo's curator of reptiles and discovered a brand-new species from Australia called the Peninsula Dragon Lizard.

The newspapers went crazy for this small blond woman handling huge pythons and lizards. To the public it was very odd to see a woman to work with such creatures! She became famous, at first for the novelty, but soon the world saw her genius. She worked closely with architects to design the zoo's reptile house, which was built in 1926 and is still used today. It was the first of its kind built specifically for the reptiles' comfort.

Joan was recognized as an expert in herpetology and published many papers on this science. Joan revealed that "the secret of a zoo is to make the animals feel at home." She used her artistic talents to make the environment look and feel like their natural habitat. On-the-job training and her special relationship with the animals made her an excellent veterinarian.

Under her care, reptiles were living longer than ever before in captivity. Her love and understanding of these reptiles led her to get to know each animal as an individual. She even kept a tame Komodo dragon as a pet.

Her chronic ill health eventually caught up with her. She would still come to work when she could, making her rounds in a wheelchair with her Komodo dragon on a leash. She died at the age of 34 in 1931, but her legacy lives on at the London Zoo.

Would feed a Komodo dragon eggs from a spoon.

Her mom was also an artist.

Used special glass in the reptile house so the animals could receive the ultraviolet sunlight.

Showed how art and painted scenery could make all animals more comfortable.

Her philosophy of creating a natural environment for the animals informs the way modern zoos are run today.

She created a perfect temperature system so all the reptiles were comfortable.

DISCOVERED THAT THE SUN IS MADE OF HYDROGEN AND HELIUM GAS.

AWARDED THE HENRY NORRIS RUSSELL PRIZE BY THE AMERICAN ASTRONOMICAL SOCIETY.

FURTHERED THE UNDERSTANDING OF STELLAR EVOLUTION.

"THERE IS NO JOY MORE INTENSE THAN THAT OF COMING UPON A FACT THAT CANNOT BE UNDERSTOOD IN TERMS OF CURRENTLY ACCEPTED IDEAS." —CECILIA PAYNE-GAPOSCHKIN

CECILIA PAYNE-GAPOSCHKIN

ASTRONOMER AND ASTROPHYSICIST

Born in England in 1900, Cecilia Payne-Gaposchkin always had a passion for learning and science. She attended Cambridge University and was inspired by a lecture on how solar eclipses relate to Einstein's Theory of General Relativity, which got her hooked on physics and astronomy.

Cambridge did not have a lot of opportunities for women and did not offer them advanced degrees. Cecilia moved from Cambridge, England, to Cambridge, Massachusetts, and started a fellowship at Harvard College Observatory, figuring out what the sun and stars are made of.

The stars could be viewed in a different way by attaching a spectroscope to a telescope. This tool allowed scientists see a rainbow of colors—the stellar spectra coming from the star. Reading the gaps in the rainbow, also known as absorption lines, revealed what types of elements were in the star.

Scientists at the time thought that stars were built like the Earth, but Cecilia proved them wrong. Her background in quantum physics gave her new insight into reading stellar spectra. She already knew that the extremely hot sun would cause atoms to ionize. Different ionization states would show up as different absorption lines on the stellar spectra. With her fresh perspective, it was now Cecilia's job to figure out the elements to which these ions could belong.

She discovered that the sun is made mostly of hydrogen and helium gas. This was so controversial that the respected astronomer Henry Russell told her it was "impossible." She completed her thesis paper with a side note saying that she was probably wrong. She turned her paper into a book, *Stellar Atmospheres*, published in 1925. Many other astronomers read the book, and in a few years the scientific community realized how right she was! Her work changed astronomy and taught scientists how to properly read stellar spectra.

Despite Cecilia's accomplishments, being a woman meant she was only recognized as a technical assistant at Harvard. Finally, in 1956, she became Harvard's first female astronomy professor. Her work has given us a better understanding of the life cycles of stars and our universe.

- HARVARD DID NOT GIVE PHDS TO WOMEN, SO SHE GOT ONE FROM RADCLIFFE COLLEGE.
- BECAME CHAIR OF THE ASTRONOMY DEPARTMENT AT HARVARD.
- WORKED ON VARIABLE STARS AND NOVAE.
- FIRST PERSON TO PROPERLY "READ" A STAR'S TEMPERATURE.
- ALSO WROTE THE BOOK STARS OF HIGH LUMINOSITY.
- WORKED AT HARVARD WITH HER HUSBAND, SERGEI GAPOSCHKIN.

CHANGED THE WAY WE UNDERSTAND EVOLUTION AND BOTANY.

PIONEER IN CORN GENETICS.

STUDIED HOW CHROMOSOMES CHANGE DURING REPRODUCTION.

"WHEN YOU HAVE THAT JOY, YOU DO THE RIGHT EXPERIMENTS. YOU LET THE MATERIAL TELL YOU WHERE TO GO, AND IT TELLS YOU AT EVERY STEP WHAT THE NEXT HAS TO BE..." — BARBARA MCCLINTOCK

BARBARA McCLINTOCK
CYTOGENETICIST

Barbara McClintock never let the expectations of others determine what she could accomplish. She was born in 1902 in Connecticut and grew up in New York City. She loved boxing, riding bikes, and playing baseball. She didn't fit in with the girls, and the boys didn't want to play with her. Against her mother's wishes, but with her father's support, she got a PhD in botany from Cornell University.

At Cornell she started her revolutionary work with corn and chromosomes.

In 1936, she started working in genetics at the University of Missouri. She was spunky, direct, and much more intelligent than many of her male peers—and this made them nervous. The dean threatened to fire her if she ever got married or if her male research partner left the university. Barbara realized they would never give a woman a full-time faculty position, so she quit in order to find her dream job.

She got down to business at a research facility in Cold Spring Harbor, New York. Barbara knew that corn was a perfect tool to explore genetics—she was fascinated by corn kernels of different colors growing on the same plant. She planted a field of corn and spent hours gazing at corn cells under a microscope.

She discovered that different colored kernels have the same genes, but they are rearranged in a different order. This meant that a gene could "jump" to a different part of a chromosome and turn on and off. The discovery of jumping genes, or "transposons," explained why there is so much variation in the world and how animals, people, and plants can evolve to react to their environment.

Excited by her discovery, Barbara gave a lecture in 1951 at the Cold Spring Harbor symposium, but no one believed her. She didn't mind, because, as she said, "When you know you're right, you don't care."

Almost 20 years later, the scientific community caught up with Barbara, and she finally received the recognition due to her. She was awarded a Nobel Prize in 1983, over 30 years after her initial discovery. The span of Barbara's work includes some of the greatest discoveries made in genetics.

AT THE UNIVERSITY OF MISSOURI SHE WAS CONSIDERED A TROUBLEMAKER FOR ALWAYS WEARING PANTS AND WORKING LATE WITH STUDENTS.

FIRST PERSON TO MAKE A COMPLETE GENETIC MAP OF CORN.

HER TECHNIQUES WERE SO ADVANCED THAT HER WORK WAS TOO CONFUSING FOR MOST SCIENTISTS AT THE TIME.

GENETICS SOCIETY OF AMERICA'S FIRST WOMAN PRESIDENT.

WAS ELECTED TO THE NATIONAL ACADEMY OF SCIENCES.

WON THE NOBEL PRIZE IN PHYSICS.

PROVED THE NUCLEAR SHELL MODEL FOR ATOMS.

GAVE US A BETTER UNDERSTANDING OF ISOTOPES.

"WHEN YOU LOVE SCIENCE, ALL YOU REALLY WANT IS TO KEEP WORKING." —MARIA GOEPPERT-MAYER

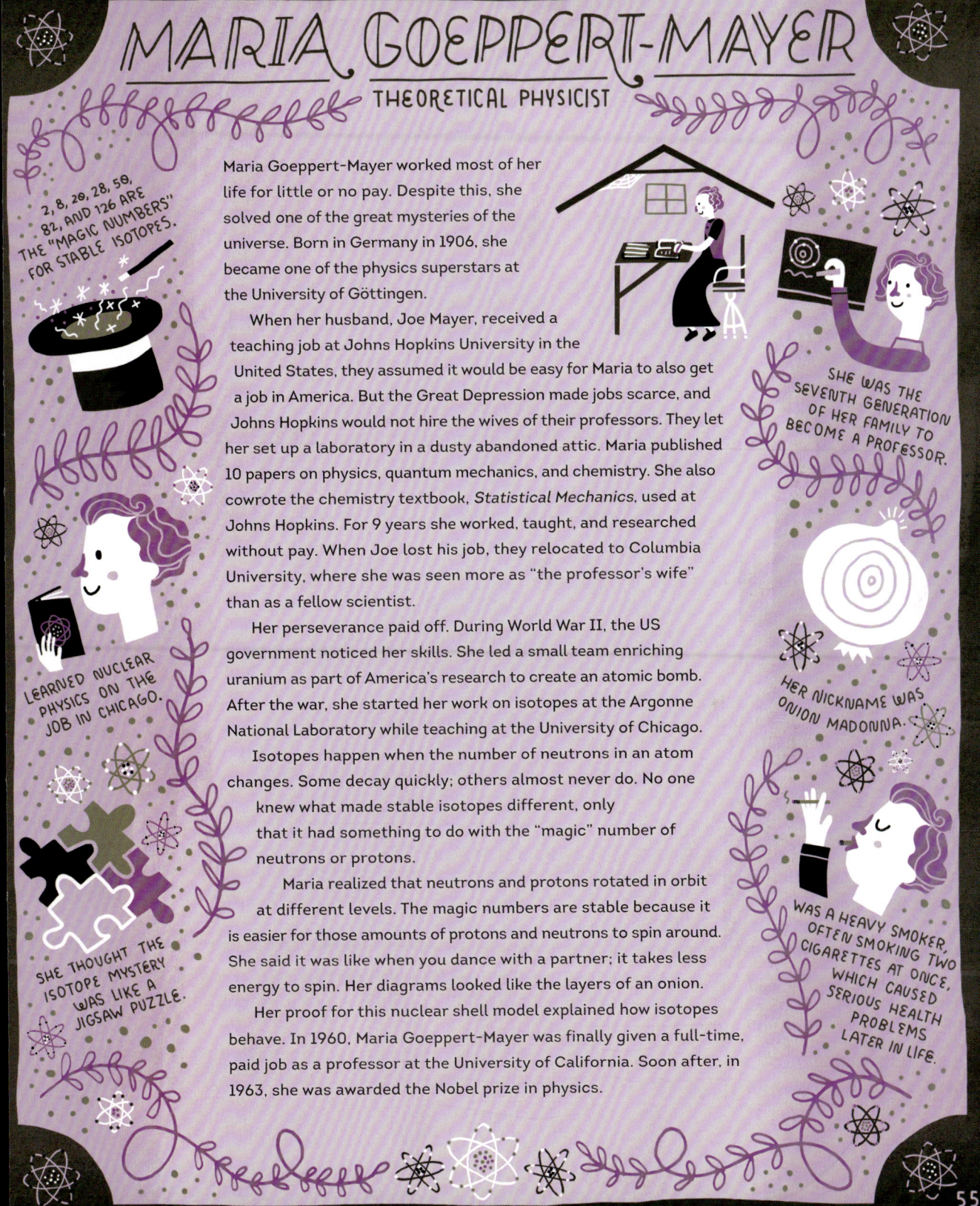

MARIA GOEPPERT-MAYER
THEORETICAL PHYSICIST

Maria Goeppert-Mayer worked most of her life for little or no pay. Despite this, she solved one of the great mysteries of the universe. Born in Germany in 1906, she became one of the physics superstars at the University of Göttingen.

When her husband, Joe Mayer, received a teaching job at Johns Hopkins University in the United States, they assumed it would be easy for Maria to also get a job in America. But the Great Depression made jobs scarce, and Johns Hopkins would not hire the wives of their professors. They let her set up a laboratory in a dusty abandoned attic. Maria published 10 papers on physics, quantum mechanics, and chemistry. She also cowrote the chemistry textbook, *Statistical Mechanics*, used at Johns Hopkins. For 9 years she worked, taught, and researched without pay. When Joe lost his job, they relocated to Columbia University, where she was seen more as "the professor's wife" than as a fellow scientist.

Her perseverance paid off. During World War II, the US government noticed her skills. She led a small team enriching uranium as part of America's research to create an atomic bomb. After the war, she started her work on isotopes at the Argonne National Laboratory while teaching at the University of Chicago.

Isotopes happen when the number of neutrons in an atom changes. Some decay quickly; others almost never do. No one knew what made stable isotopes different, only that it had something to do with the "magic" number of neutrons or protons.

Maria realized that neutrons and protons rotated in orbit at different levels. The magic numbers are stable because it is easier for those amounts of protons and neutrons to spin around. She said it was like when you dance with a partner; it takes less energy to spin. Her diagrams looked like the layers of an onion.

Her proof for this nuclear shell model explained how isotopes behave. In 1960, Maria Goeppert-Mayer was finally given a full-time, paid job as a professor at the University of California. Soon after, in 1963, she was awarded the Nobel prize in physics.

2, 8, 20, 28, 50, 82, AND 126 ARE THE "MAGIC NUMBERS" FOR STABLE ISOTOPES.

LEARNED NUCLEAR PHYSICS ON THE JOB IN CHICAGO.

SHE THOUGHT THE ISOTOPE MYSTERY WAS LIKE A JIGSAW PUZZLE.

SHE WAS THE SEVENTH GENERATION OF HER FAMILY TO BECOME A PROFESSOR.

HER NICKNAME WAS ONION MADONNA.

WAS A HEAVY SMOKER, OFTEN SMOKING TWO CIGARETTES AT ONCE, WHICH CAUSED SERIOUS HEALTH PROBLEMS LATER IN LIFE.

INVENTED THE FIRST COMPILER, FOREVER CHANGING HOW WE USE COMPUTERS.

CREATED COBOL, THE FIRST COMPLEX COMPUTER LANGUAGE.

PIONEERED THE STANDARDS FOR TESTING COMPUTER SYSTEMS.

"PEOPLE ARE ALLERGIC TO CHANGE. YOU HAVE TO GET OUT AND SELL THE IDEA." — GRACE HOPPER

GRACE HOPPER
NAVY ADMIRAL and COMPUTER SCIENTIST

Grace Hopper was a Navy admiral and a relentless trailblazer, recognized as the mother of computer programming. She was born in New York City in 1906 and earned a PhD in mathematics at Yale in 1934. Grace was working as a math professor at Vassar College when the United States entered World War II. In 1943, Grace quit her job to join the Women Accepted for Volunteer Emergency Service (WAVES). Even though she was too small to meet the physical requirements, her mathematical mind was exactly what the country needed.

The Navy assigned her to Harvard University to program one of the first-ever electronic computers. When Grace saw the Mark I, she thought, "Gee, that's the prettiest gadget I ever saw." She was second in command to Howard Aiken, one of the original designers of the machine.

Back then calculations were done by a large group of people. This new computer would be able to solve equations that were too complicated for that old system. Grace's team used the Mark I to solve important problems for the war effort, including the implosion equation for the Manhattan Project.

After the war, Grace joined the private sector. At the time, programmers needed the skills that came with an advanced degree in mathematics and used binary code to program. Grace Hopper thought it would be easier to just "talk" to a computer in English. Everyone thought Grace was nuts, but she proved them wrong when she invented the first compiler. This led her to create COBOL, the first universal computer language. Thanks to Grace, just about anyone can learn to code!

Grace returned to the Navy in 1967. Even after she retired as the oldest person on active duty (just a few months short of turning 80), she continued to lecture, consult, and teach—always reminding the world that "the most damaging phrase in the language is 'we've always done it this way.'"

RECEIVED THE DEFENSE DISTINGUISHED SERVICE MEDAL.

HAD A BACKWARD CLOCK IN HER OFFICE TO REMIND HER THAT THINGS DON'T HAVE TO WORK JUST ONE WAY.

APPEARED ON THE LATE SHOW WITH DAVID LETTERMAN AND 60 MINUTES.

COINED THE TERM "DEBUGGING" WHEN A MOTH GOT CAUGHT IN THE COMPUTER.

HER GREAT-GRANDFATHER WAS ALSO IN THE NAVY.

THE MARK I COMPUTER WAS 51 FEET WIDE.

HAD A JOLLY ROGER PIRATE FLAG ON HER DESK BECAUSE SHE WAS RELENTLESS IN GETTING WHAT HER TEAM NEEDED.

11.8 INCHES — FAMOUS FOR HER CUT WIRES SHOWING THE DISTANCE THAT ELECTRICITY TRAVELS IN A NANOSECOND.

TAUGHT THE WORLD ABOUT THE OCEAN'S ECOSYSTEMS.

INSPIRED THE US ENVIRONMENTAL PROTECTION AGENCY.

WROTE AWARD-WINNING BOOKS: THE SEA AROUND US, THE EDGE OF THE SEA, UNDER THE SEA WIND, & SILENT SPRING.

"THE HUMAN RACE IS CHALLENGED MORE THAN EVER BEFORE TO DEMONSTRATE OUR MASTERY, NOT OVER NATURE BUT OF OURSELVES." — RACHEL CARSON

RACHEL CARSON

MARINE BIOLOGIST, CONSERVATIONIST, AND AUTHOR

From an early age, Rachel Carson could always be found looking at birds, bugs, and fish. She was born in 1907 and grew up on a Pennsylvania farm. She got her master's in zoology at Johns Hopkins University, but when her father died, Rachel decided against getting a doctorate so she could instead support her family. She became the second woman to work at the US Bureau of Fisheries, writing radio scripts about sea creatures. When she wasn't at her government job, she did personal writing about wildlife.

Rachel's poetic writing allowed her to reach people in all walks of life. Her first book, *Under the Sea Wind*, got little attention, but her next book, *The Sea Around Us*, became a sensation! She won the National Book Award and quit her job to write *The Edge of the Sea*.

In the 1950s, the US government and private industry started to blindly overuse the pesticide DDT. We now know that DDT is highly toxic and that large doses can cause liver damage and seizures. DDT was being used everywhere, from the bug spray you'd use at picnics to all of our crops—but it killed more than just pests.

Rachel Carson received a letter from an old friend, Olga Huckins, when a plane spraying DDT killed all the song birds in her sanctuary. This inspired Rachel to research and write her greatest book, *Silent Spring*. Rachel's research found that DDT was poisoning livestock, killing fish, fatally weakening birds' eggs, and wreaking havoc on the ecosystem.

She wrote the book while battling cancer, and she needed to constantly defend her findings. Chemical companies slandered her work, but Rachel would not be bullied, and the truth about DDT became public. She even spoke in front of the US Senate.

She died in 1964, 2 years after *Silent Spring* was published. The book raised awareness—and action would follow. Rachel's work was directly responsible for the creation of the US Environmental Protection Agency and inspired the environmental movement around the world.

WROTE A BOOK ABOUT BIRDS WHEN SHE WAS EIGHT.

WAS PUBLISHED IN A CHILDREN'S MAGAZINE WHEN SHE WAS 11.

THE NATIONAL ENVIRONMENTAL POLICY ACT (NEPA) WAS PASSED IN RESPONSE TO SILENT SPRING.

CHEMICAL COMPANIES SPENT NEARLY $250,000 ON A SMEAR CAMPAIGN TO DISCREDIT RACHEL.

HER GOVERNMENT RADIO PROGRAM ABOUT FISH WAS CALLED "ROMANCE UNDER THE WATERS."

SILENT SPRING WAS EXHAUSTIVELY RESEARCHED: IT HAS A 55-PAGE BIBLIOGRAPHY.

AWARDED THE PRESIDENTIAL MEDAL OF FREEDOM IN 1980.

LAB TOOLS

Problem solving requires testing and experimentation, and having the right equipment can make or break your research. These women did their work anywhere they could—from dusty attics to tiny sheds—or, once they gained recognition and respect, in state-of-the-art labs.

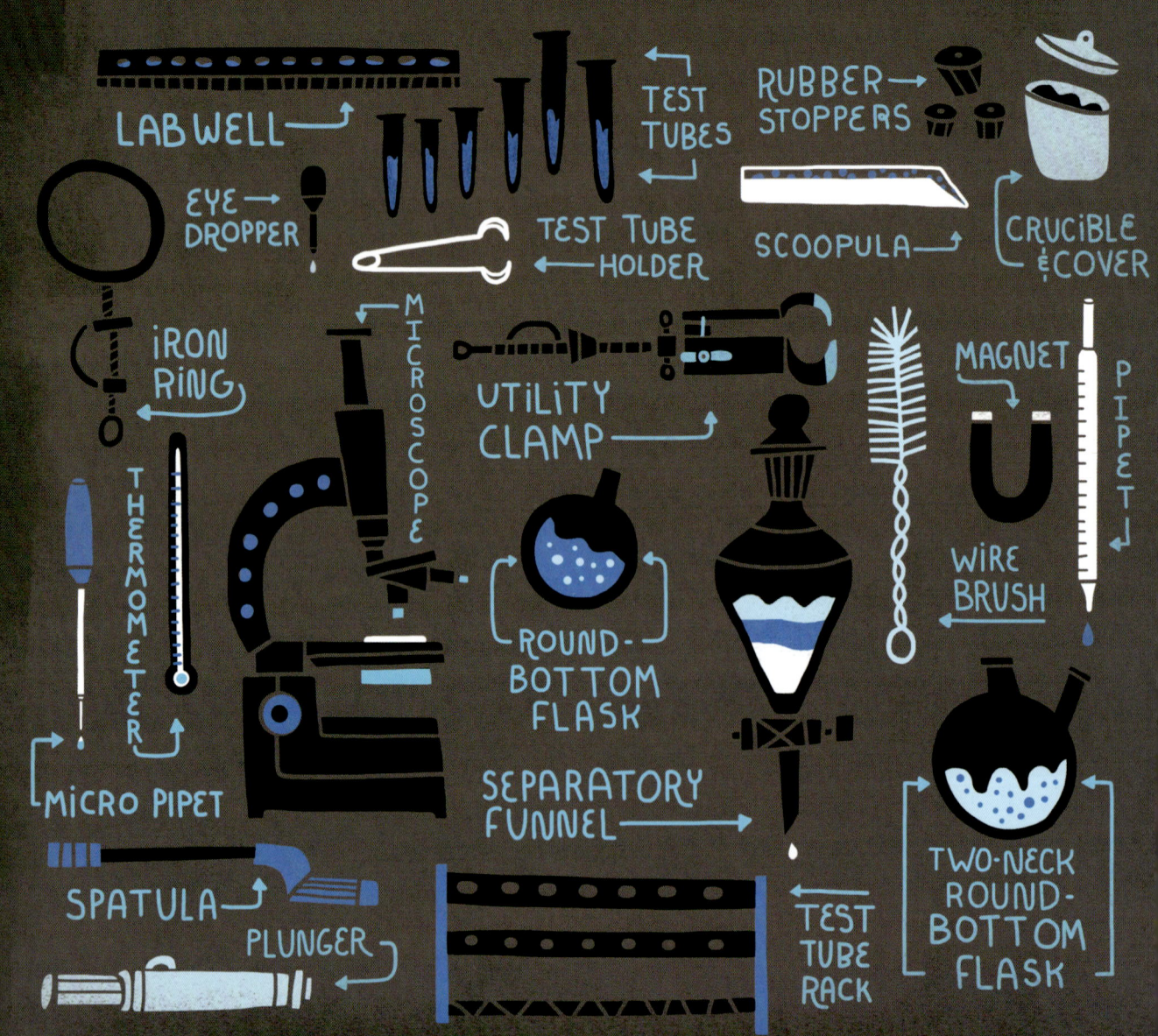

- LAB WELL
- TEST TUBES
- RUBBER STOPPERS
- EYE DROPPER
- TEST TUBE HOLDER
- SCOOPULA
- CRUCIBLE & COVER
- IRON RING
- MICROSCOPE
- UTILITY CLAMP
- MAGNET
- PIPET
- THERMOMETER
- ROUND-BOTTOM FLASK
- WIRE BRUSH
- MICRO PIPET
- SEPARATORY FUNNEL
- SPATULA
- PLUNGER
- TEST TUBE RACK
- TWO-NECK ROUND-BOTTOM FLASK

60

- BÜCHNER FUNNEL
- CORK STOPPERS
- PETRI DISH
- SLIDE
- RULER
- TWO-NECK FLASK
- TONGS
- LAB BURNER
- SAFETY GOGGLES
- WATCH GLASS
- BURETTE
- WIRE GAUZE
- WASH BOTTLE
- FLORENCE FLASK
- SCALPEL
- FUNNEL
- MAGNIFYING GLASS
- EVAPORATING DISH
- GRADUATED CYLINDER
- LIGHTER
- PIPE-STEM TRIANGLE
- 24-WELL PLATE
- POWDER FUNNEL
- BUNSEN BURNER
- FORCEPS
- FILE
- MORTAR & PESTLE
- EXTENSION CLAMP
- ERLENMEYER FLASK
- BEAKER

61

- Awarded the Nobel Prize in Physiology or Medicine.
- Discovered the nerve growth factor.
- Was an Italian senator for life.

"Above all, don't fear difficult moments. The best comes from them." —Rita Levi-Montalcini

RITA LEVI-MONTALCINI

NEUROLOGIST AND ITALIAN SENATOR

Rita Levi-Montalcini never let her circumstance keep her from science. She was born in 1909 in Italy to a well-to-do Jewish family. Her father expected her to become a proper lady and marry well, but she hated finishing school and was determined to become a doctor.

Though Rita graduated summa cum laude from medical school in 1936, she had no real job prospects. Italy was one of the Axis powers in World War II, and in 1938, anti-Semitic laws forbade Jewish people to practice medicine. But nothing could keep Rita from pursuing her dreams.

She created a makeshift laboratory in her bedroom and started her research. She borrowed eggs from farmers and used sewing needles to dissect the nervous systems of embryonic chicks. She wanted to know why and how nerve cells developed. By severing the limbs of the chick embryo, she accurately documented how the motor neurons began to grow and then die. This work laid the foundation for her entire career.

When the war ended, Rita reentered the formal scientific world, already well into her research. She was asked to come to Washington University in Saint Louis, Missouri, for one semester, which turned into 30 years of teaching and research.

While learning how to grow tissues in a glass dish, Rita observed that a tumor sample was affecting the embryonic cells in the same dish. The nerves started to grow very quickly—but why? By experimenting with snake venom, tumors, and finally mouse saliva, she discovered nerve growth factor (NGF), a protein that regulates nerve growth and keeps our neurons healthy. This was a very important finding for understanding and fighting diseases.

Rita received the 1986 Nobel prize in physiology or medicine. When asked if she was bitter about how the Italian government treated her during the war, she said, "If I had not been discriminated against or had not suffered persecution, I would never have received the Nobel prize." She went on to become a senator-for-life in the Italian government, where she fought for civic equality and promoted the sciences.

She was appointed to the Pontifical Academy of Sciences by Pope Paul VI—

And shook the Pope's hand instead of kissing it.

Snuck lab mice onto a plane for her research.

Worked until she died at the age 103.

Once lectured in a pressed night gown when her luggage was lost.

She did important research on human mast cells and their relation to NGF.

Shared the Nobel prize with her lab partner and collaborator Stanley Cohen.

DISCOVERED THE STRUCTURE OF PENICILLIN, VITAMIN B12, AND INSULIN.

SHE INVENTED TECHNIQUES TO USE X-RAY CRYSTALLOGRAPHY TO MAP COMPLEX MOLECULES.

WON A NOBEL PRIZE IN CHEMISTRY AND THE ORDER OF MERIT.

— "I WAS CAPTURED FOR LIFE BY CHEMISTRY AND BY CRYSTALS." — DOROTHY HODGKIN —

DOROTHY HODGKIN

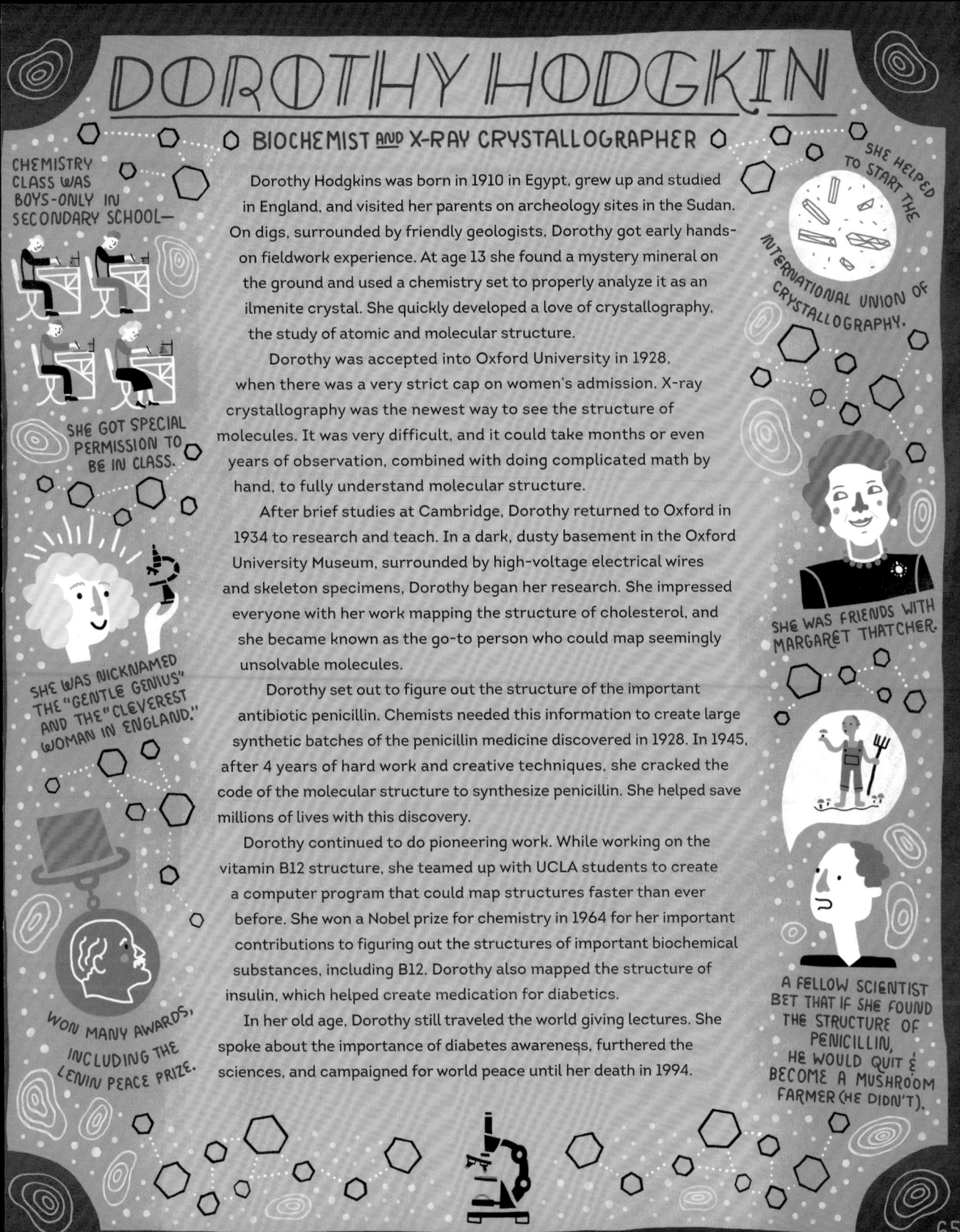

BIOCHEMIST and X-RAY CRYSTALLOGRAPHER

Dorothy Hodgkins was born in 1910 in Egypt, grew up and studied in England, and visited her parents on archeology sites in the Sudan. On digs, surrounded by friendly geologists, Dorothy got early hands-on fieldwork experience. At age 13 she found a mystery mineral on the ground and used a chemistry set to properly analyze it as an ilmenite crystal. She quickly developed a love of crystallography, the study of atomic and molecular structure.

Dorothy was accepted into Oxford University in 1928, when there was a very strict cap on women's admission. X-ray crystallography was the newest way to see the structure of molecules. It was very difficult, and it could take months or even years of observation, combined with doing complicated math by hand, to fully understand molecular structure.

After brief studies at Cambridge, Dorothy returned to Oxford in 1934 to research and teach. In a dark, dusty basement in the Oxford University Museum, surrounded by high-voltage electrical wires and skeleton specimens, Dorothy began her research. She impressed everyone with her work mapping the structure of cholesterol, and she became known as the go-to person who could map seemingly unsolvable molecules.

Dorothy set out to figure out the structure of the important antibiotic penicillin. Chemists needed this information to create large synthetic batches of the penicillin medicine discovered in 1928. In 1945, after 4 years of hard work and creative techniques, she cracked the code of the molecular structure to synthesize penicillin. She helped save millions of lives with this discovery.

Dorothy continued to do pioneering work. While working on the vitamin B12 structure, she teamed up with UCLA students to create a computer program that could map structures faster than ever before. She won a Nobel prize for chemistry in 1964 for her important contributions to figuring out the structures of important biochemical substances, including B12. Dorothy also mapped the structure of insulin, which helped create medication for diabetics.

In her old age, Dorothy still traveled the world giving lectures. She spoke about the importance of diabetes awareness, furthered the sciences, and campaigned for world peace until her death in 1994.

CHEMISTRY CLASS WAS BOYS-ONLY IN SECONDARY SCHOOL—

SHE GOT SPECIAL PERMISSION TO BE IN CLASS.

SHE WAS NICKNAMED THE "GENTLE GENIUS" AND THE "CLEVEREST WOMAN IN ENGLAND."

WON MANY AWARDS, INCLUDING THE LENIN PEACE PRIZE.

SHE HELPED TO START THE INTERNATIONAL UNION OF CRYSTALLOGRAPHY.

SHE WAS FRIENDS WITH MARGARET THATCHER.

A FELLOW SCIENTIST BET THAT IF SHE FOUND THE STRUCTURE OF PENICILLIN, HE WOULD QUIT & BECOME A MUSHROOM FARMER (HE DIDN'T).

DISPROVED A "LAW OF CONSERVATION OF PARITY."

AWARDED THE MEDAL OF SCIENCE.

HELPED DEVELOP THE FUEL FOR THE ATOMIC BOMB.

"THE MAIN STUMBLING BLOCK IN THE WAY OF ANY PROGRESS IS AND ALWAYS HAS BEEN UNIMPEACHABLE TRADITION." — CHIEN-SHIUNG WU

CHIEN-SHIUNG WU

EXPERIMENTAL PHYSICIST

Chien-Shiung Wu was born in China in 1912, when not all women were expected to become educated. Chien-Shiung Wu's father was a pioneer for women's rights and started the first school in their town for girls. Her family always gave her the support to attend the best schools no matter the distance or the price. In 1936, Chien-Shiung Wu headed to the United States to continue her studies in experimental physics.

After graduating with a PhD from the University of California in 1940, Chien-Shiung Wu became a professor at Princeton University and Smith College. Wu was known to be demanding with her assignments but pushed her students to be their best, and they loved her for it.

World War II was fought and won with science, and in 1944, Chien-Shiung Wu was recruited to Columbia University to work on the Manhattan Project. She helped develop a way to enrich uranium into the isotopes needed to fuel the atomic bomb. She helped to develop radiation detectors for the project as well.

After the war, Chien-Shiung Wu stayed at Columbia to start her work on beta decay. The theory called the "law of conservation of parity" predicted that radioactive atoms decay in a symmetrical way. But a new particle was discovered, called the K-meson, that didn't follow the rules. No one actually observed this wonky particle until Chien-Shiung Wu did. She labored night and day, skipping vacations and forcing her assistants to work weekends with her.

With determination and a very strong magnet, Chien-Shiung Wu observed that the electrons of these atoms broke away asymmetrically. She disproved the "law of conservation of parity" and changed the practice of physics forever.

She published a book, *Beta Decay*, and was given many awards and honors. She continued to research and lecture around the globe into her old age.

- WON THE 1975 NATIONAL MEDAL OF SCIENCE.
- WHILE BEING INTERVIEWED FOR THE TOP-SECRET MANHATTAN PROJECT, SHE ALREADY KNEW WHAT THEY WERE WORKING ON JUST BY LOOKING AT AN EQUATION LEFT ON A BLACKBOARD.
- NICKNAMED "FIRST LADY OF PHYSICS."
- APS FIRST WOMAN ELECTED A FELLOW TO THE AMERICAN PHYSICAL SOCIETY.
- RESEARCHED SICKLE CELL DISEASE.
- HER NAME TRANSLATES AS "COURAGEOUS HERO."

"[MY FATHER] MADE ME UNDERSTAND THAT I MUST MAKE MY OWN DECISIONS, MOLD MY OWN CHARACTER, THINK MY OWN THOUGHTS." —HEDY LAMARR

HEDY LAMARR
INVENTOR AND FILM ACTRESS

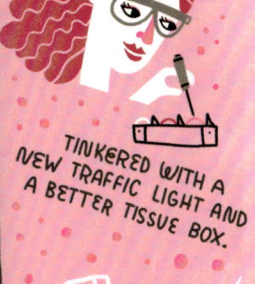

You may already know that Hedy Lamarr was an actress during Hollywood's Golden Age who was called "the most beautiful woman in the world." But unbeknownst to most, she was also a genius inventor!

Hedy was born Hedwig Eva Maria Kiesler in 1914 in Vienna, Austria. She dreamed of being an actress—and made it a reality. When her controlling millionaire husband, Fritz Mandl, wanted to put an end to her acting career, she left him and fled to Paris and later, London. There she met Louis B. Mayer, a big-time film producer who later gave her an acting contract from MGM and a new name.

Hedy also had a secret workshop where she tinkered with inventions. During World War II, the National Inventors Council asked civilians to submit ideas. Hedy identified a problem she thought she could fix: the US Navy's radio-guided torpedoes were easy to signal jam, which caused them to go off course.

At a dinner party, she met George Antheil, an avant-garde composer. Together they realized a radio signal could change frequencies using the same technology a player piano uses to change notes. The signal would be impossible to jam. Hedy was so excited she wrote her number in lipstick on his car window and immediately got to work. Working together, they developed the frequency-hopping spread spectrum (FHSS). She received a patent in 1942, but the US military shelved her idea. Disheartened but still patriotic, Hedy used her fame to raise millions of dollars in war bonds. It wasn't until the Cuban missile crisis in 1962 that the military realized FHSS was a goldmine. Hedy's technology was used to control torpedoes and communication. FHSS is especially useful for communication between multiple electronic devices—it's the foundation for the technology we now use every day with our smartphones, GPS, Wi-Fi, and Bluetooth devices.

Although the patent had expired by the time FHSS was used, Hedy won many awards while she was alive and was inducted into the National Inventors Hall of Fame in 2014, 14 years after her death.

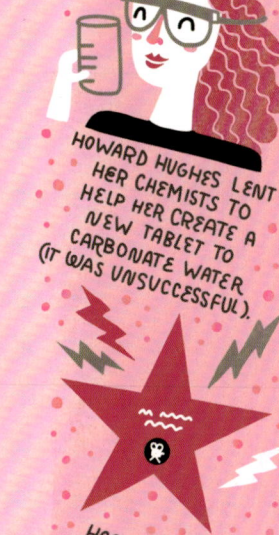

- TINKERED WITH A NEW TRAFFIC LIGHT AND A BETTER TISSUE BOX.
- RECEIVED THE ELECTRONIC FRONTIER FOUNDATION PIONEER AWARD IN 1997.
- STARRED IN FILMS WITH CLARK GABLE, SPENCER TRACY, AND JIMMY STEWART.
- HOWARD HUGHES LENT HER CHEMISTS TO HELP HER CREATE A NEW TABLET TO CARBONATE WATER. (IT WAS UNSUCCESSFUL).
- HAS A STAR IN THE HOLLYWOOD WALK OF FAME.
- EX-HUSBAND, FRITZ MANDL, WAS A WEAPONS MANUFACTURER;
- HEDY LEARNED TRADE SECRETS FROM OVERHEARING HIS DINNER CONVERSATIONS.

Her work won the Supreme Court case Brown v. Board of Education.

Co founded the Northside Center for Child Development in Harlem, New York City.

Her doll test and coloring test proved that segregation hurts children.

"What did it mean that all these children were in one place?... They're isolated from whites, and they can never learn that they're just as good as whites... You have to get these children desegregated." —Mamie Phipps Clark

MAMIE PHIPPS CLARK
PSYCHOLOGIST AND CIVIL RIGHTS ACTIVIST

Slavery in America was abolished in 1865, but even though African-Americans were free in name, they would not gain full equality under the law until the Fair Housing Act of 1968. For over 100 years, black Americans would be denied the right to vote, the right to a proper education, and the right to exist in certain places.

Mamie Phipps Clark was born in 1917 in Arkansas. Racial segregation in the south meant that Mamie was not allowed in stores owned by white people and had to attend poorly funded black-only schools. Even so, Mamie had a happy childhood filled with love and learning.

Mamie met her husband and future partner in psychology, Kenneth Clark, at Howard University. Mamie's master's thesis was "The Development of Consciousness of Self in Negro Pre-School Children." She used a picture test to prove that race is an integral part of a child's identity. Mamie realized she could use psychology to prove that segregation was wrong.

Mamie got her PhD from Columbia University in 1943. Mamie and Kenneth eventually started their own practice, providing psychological help to families in New York's black community.

Together the Clarks started the Doll Experiment. They traveled the country and compared the responses of children in segregated and integrated schools. They gave the children identical black and white dolls and asked, "Which doll do you want to play with? Is this doll pretty? Is this doll nice?"

It became clear that black children identified with the black doll, but the children in segregated schools said the black doll was ugly and bad and thought they themselves were also bad. Mamie and Kenneth had tangible proof that segregation damaged children and caused self-hate. This study was used in the 1954 Supreme Court case *Brown v. Board of Education*, which ended segregation in public schools.

Although laws have changed, the effects of segregation are still felt in America. This wound will need pioneers and activists like Mamie Phipps Clark to fully heal. Together, we need to work hard to continue to fight the injustice that persists today.

KENNETH CLARK

SHE GRADUATED MAGNA CUM LAUDE FROM HOWARD UNIVERSITY.

CHOSE HER CAREER BECAUSE SHE ALWAYS WANTED TO WORK WITH CHILDREN.

WORKED AS A COUNSELOR FOR HOMELESS AFRICAN-AMERICAN GIRLS AT THE RIVERDALE HOME IN NEW YORK.

THE COLORING TEST ALSO PROVED THAT SEGREGATION WARPS SELF-WORTH.

WAS THE SECOND AFRICAN-AMERICAN (HER HUSBAND WAS THE FIRST) TO RECEIVE A PH.D FROM COLUMBIA UNIVERSITY.

WAS THE DIRECTOR OF NORTHSIDE CENTER FOR CHILD DEVELOPMENT FROM 1946 UNTIL SHE RETIRED IN 1979.

EQUAL RIGHTS

EQUAL RIGHTS

HAS SAVED THOUSANDS OF LIVES WITH THE DRUGS SHE DEVELOPED.

HELPED TO DEVELOP DRUGS TO FIGHT CANCER, AIDS, HERPES, AND MANY MORE DISEASES.

FIRST WOMAN INDUCTED INTO THE NATIONAL INVENTORS HALL OF FAME.

"DON'T LET OTHERS DISCOURAGE YOU OR TELL YOU THAT YOU CAN'T DO IT. IN MY DAY I WAS TOLD WOMEN DIDN'T GO INTO CHEMISTRY. I SAW NO REASON WHY WE COULDN'T." —GERTRUDE ELION

GERTRUDE ELION
PHARMACOLOGIST AND BIOCHEMIST

Gertrude Elion was born in 1918 and grew up in the Bronx in New York City. She was a great student who loved all of her subjects in high school and graduated early at age 15. She didn't know what her career would be until her grandfather died of cancer. She decided to dedicate her life to fighting the disease.

During the Great Depression, universities prioritized hiring men. Gertrude graduated with high honors from Hunter College, but graduate schools were offering no financial aid to women, and chemistry jobs were scarce. Finally, after many odd jobs and one cash-strapped year in NYU's graduate program, she found a home for her cancer research at the Burroughs Wellcome pharmaceutical company.

That group broke away from the usual trial-and-error way of developing drugs. With George Hitchings, she studied the difference between healthy and abnormal cells and how abnormal cells reproduce so they could create drugs that destroy only unhealthy cells. Gertrude was tasked with studying the nucleic acids in DNA and how they can be used to stop tumors from spreading.

She started working toward finishing her PhD part time at night. Her school demanded that she attend full time and quit her job, but she loved her work so much that she quit the PhD program instead. It was the right choice. Gertrude went on to create many different medications that saved thousands of lives. In 1950, she created 2 drugs to treat leukemia, which began a new era of cancer research.

Gertrude continued to work with many different diseases. Another major breakthrough came in 1978, when she created a way for antivirals to accurately target a virus without harming healthy cells. A resulting drug is used to treat herpes and has been the basis for many other antivirals.

Gertrude's drug research saved thousands of lives and made tremendous advances in drug treatment. When asked her favorite achievement, she responded, "I don't discriminate among my children."

- WON THE NOBEL PRIZE IN PHYSIOLOGY OR MEDICINE IN 1988.
- SHE CREATED MEDICATION FOR GOUT AND SHINGLES.
- BECAME A DEPARTMENT HEAD AT BURROUGHS WELLCOME.
- AFTER RETIRING, BECAME A RESEARCH PROFESSOR AT DUKE UNIVERSITY.
- HER FIRST FULL-TIME CHEMISTRY JOB WAS TESTING PICKLES FOR GROCERY STORES.
- HER HERPES DRUG LED TO AZT, WHICH FIGHTS AIDS.
- WAS PRESIDENT OF THE AMERICAN ASSOCIATION FOR CANCER RESEARCH AND PARTICIPATED IN MANY MORE CANCER ORGANIZATIONS.

PHYSICIST, SPACE SCIENTIST, AND NASA MATHEMATICIAN.

$$v_e = \sqrt{\frac{2GM}{r}}$$

WON THE NASA LUNAR ORBITER AWARD & THE NASA SPECIAL ACHIEVEMENT AWARD.

CALCULATED THE FLIGHT PATH FOR THE FIRST MANNED MISSION TO THE MOON.

HAS WORKED ON NASA'S MERCURY MISSIONS, SPACE SHUTTLES, AND PLANS FOR THE MISSION TO MARS.

"[THE OTHER WOMEN] DIDN'T ASK QUESTIONS OR TAKE THE TASK ANY FURTHER. I ASKED QUESTIONS; I WANTED TO KNOW WHY. THEY GOT USED TO ME ASKING QUESTIONS AND BEING THE ONLY WOMAN THERE." —KATHERINE JOHNSON

KATHERINE JOHNSON

PHYSICIST AND MATHEMATICIAN

Katherine Johnson was born in 1918 in West Virginia and always had a love of learning and math. She excelled in school and enrolled at West Virginia State College when she was only 15 years old.

Katherine assumed she was going to become a math teacher or a nurse, like other women she knew, until she got to college and met her professor, the famous mathematician W. W. Schieffelin Claytor. He inspired Katherine to become a research mathematician and helped her pick out the classes she would need to achieve this goal.

When she was 18, Katherine graduated college. It was the height of the Great Depression and jobs were scarce, so she fell back on teaching in high school. In the 1950s, NASA began to have more openings for African-American female computers. Katherine applied and got a job!

Katherine wanted to know the in and outs of what she was working on. She was not allowed in meetings, so she asked if it was against the law for a woman to be in one. Her boldness and curiosity paid off, and she was included. Calculating flight paths involved complicated geometry equations, and Katherine was extremely good at these. She was pulled into working on the 1961 manned Mercury mission and successfully calculated the launch window.

Her skill in mathematics was on point; she quickly became a leader in calculating trajectory, making her an essential part of the team that calculated the path for the first manned mission to the moon in 1969. She did most of the calculations on the project and was also in charge of checking the math of the brand-new mechanical computers at NASA. The math had to be perfect if the Apollo team was to return to Earth safely. The Apollo mission was a success, and her crucial contributions made it possible!

Katherine later worked on lots of important NASA projects, including the space shuttle program and plans for the mission to Mars. Her work has helped astronauts visit the stars and come safely back to earth. She retired after 33 years of service in 1986.

As a little girl, she loved numbers and would count everything she could find.

In 2015 she won the Presidential Medal of Freedom at age 97.

Coauthored 26 scientific papers.

Majored in math and French in college.

The moon & the Apollo shuttle move at different speeds—her calculations ensured they would meet.

Was the 1997 Mathematician of the Year.

Helped write the first textbook about space travel.

Received an honorary Doctor of Law degree from the State University of New York.

COFOUNDED THE AMERICAN SOCIETY OF CLINICAL ONCOLOGY.

HER NEW CHEMO TECHNIQUES SAVED MILLIONS OF LIVES.

DEVELOPED NEW WAYS TO TEST CHEMO DRUGS AND TO TREAT HARD-TO-REACH TUMORS.

"NOT ONLY WAS [JANE WRIGHT'S] WORK SCIENTIFIC, BUT IT WAS VISIONARY FOR THE WHOLE SCIENCE OF ONCOLOGY." —DR. SANDRA SWAIN, THE NEW YORK TIMES

JANE COOKE WRIGHT
ONCOLOGIST

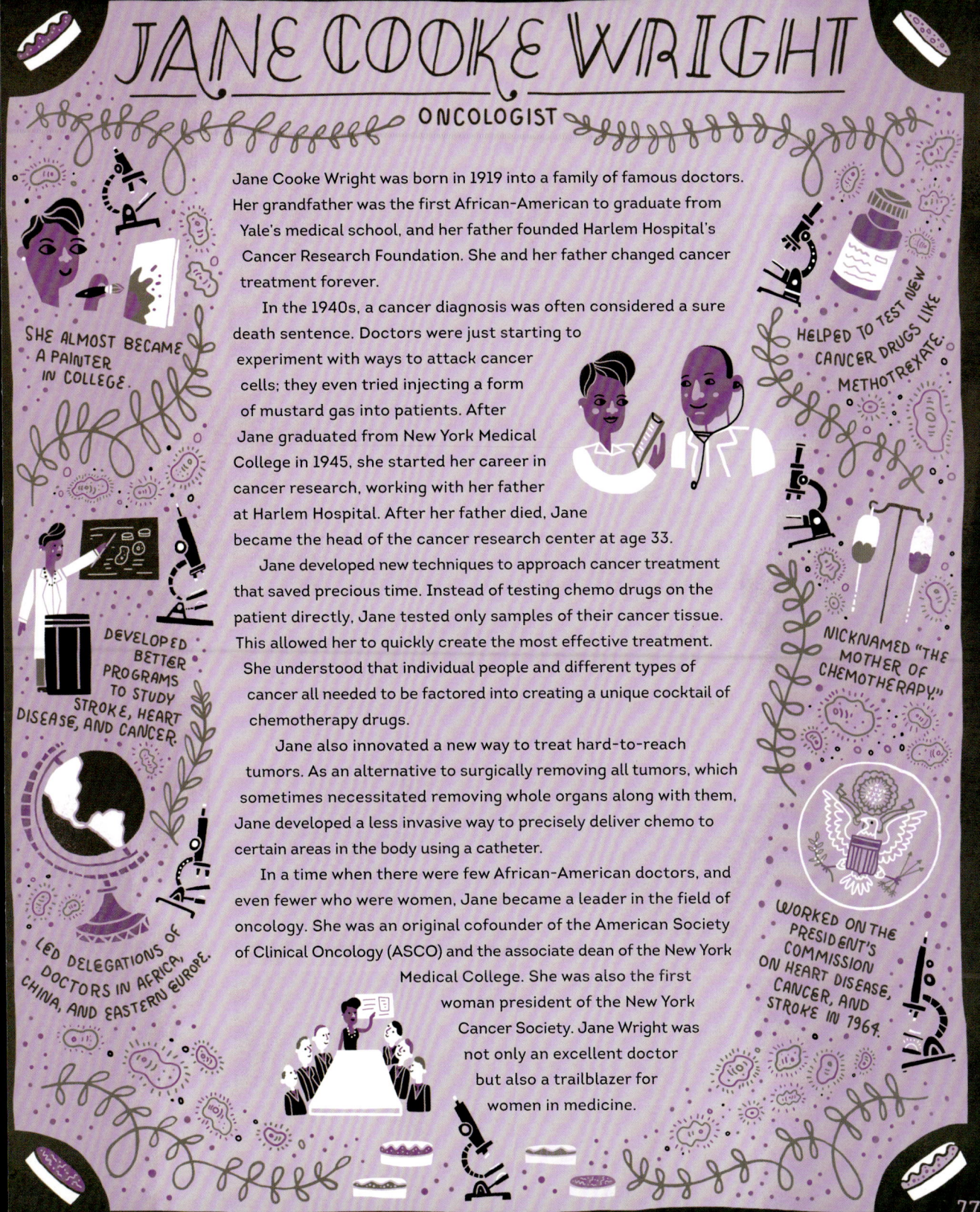

Jane Cooke Wright was born in 1919 into a family of famous doctors. Her grandfather was the first African-American to graduate from Yale's medical school, and her father founded Harlem Hospital's Cancer Research Foundation. She and her father changed cancer treatment forever.

In the 1940s, a cancer diagnosis was often considered a sure death sentence. Doctors were just starting to experiment with ways to attack cancer cells; they even tried injecting a form of mustard gas into patients. After Jane graduated from New York Medical College in 1945, she started her career in cancer research, working with her father at Harlem Hospital. After her father died, Jane became the head of the cancer research center at age 33.

Jane developed new techniques to approach cancer treatment that saved precious time. Instead of testing chemo drugs on the patient directly, Jane tested only samples of their cancer tissue. This allowed her to quickly create the most effective treatment. She understood that individual people and different types of cancer all needed to be factored into creating a unique cocktail of chemotherapy drugs.

Jane also innovated a new way to treat hard-to-reach tumors. As an alternative to surgically removing all tumors, which sometimes necessitated removing whole organs along with them, Jane developed a less invasive way to precisely deliver chemo to certain areas in the body using a catheter.

In a time when there were few African-American doctors, and even fewer who were women, Jane became a leader in the field of oncology. She was an original cofounder of the American Society of Clinical Oncology (ASCO) and the associate dean of the New York Medical College. She was also the first woman president of the New York Cancer Society. Jane Wright was not only an excellent doctor but also a trailblazer for women in medicine.

- SHE ALMOST BECAME A PAINTER IN COLLEGE.
- DEVELOPED BETTER PROGRAMS TO STUDY STROKE, HEART DISEASE, AND CANCER.
- LED DELEGATIONS OF DOCTORS IN AFRICA, CHINA, AND EASTERN EUROPE.
- HELPED TO TEST NEW CANCER DRUGS LIKE METHOTREXATE.
- NICKNAMED "THE MOTHER OF CHEMOTHERAPY."
- WORKED ON THE PRESIDENT'S COMMISSION ON HEART DISEASE, CANCER, AND STROKE IN 1964.

DID CRITICAL WORK ON MOLECULAR STRUCTURES OF DNA, RNA, VIRUSES, COAL, & GRAPHITE.

DISCOVERED THE DNA DOUBLE HELIX.

PIONEERED RESEARCH ON THE TOBACCO MOSAIC VIRUS & POLIO.

"SCIENCE AND EVERYDAY LIFE CANNOT AND SHOULD NOT BE SEPARATED." —ROSALIND FRANKLIN

ROSALIND FRANKLIN
CHEMIST AND X-RAY CRYSTALLOGRAPHER

Rosalind Franklin was born in 1920 in London. Her father wanted her to do work that he considered fit for a lady; he disapproved of women going to university. The women in Rosalind's family helped her stand up to her father. She went on to earn a PhD in physical chemistry from Cambridge University.

The big question of the day was this: what is the shape of DNA? Scientists knew that DNA formed the building blocks for the body, but they had no proof of what it really looked like. Rosalind Franklin was one of the scientists at King's College who were on the case.

Rosalind spent hours and hours using an X-ray on the delicate fibers of DNA. She captured the famous photo that proved DNA is a double helix.

Meanwhile, 2 scientists, James Watson and Francis Crick, were also trying to figure out the structure of DNA. They snuck a peek at Rosalind's work, without her permission, and used her findings to publish their own work without giving her any credit. As a result, she was overlooked. Rosalind left the toxic work environment of King's College and continued her research. She went on to a top research lab and started doing interesting research with the tobacco mosaic virus and the polio virus.

Unfortunately, Rosalind was diagnosed with terminal cancer, probably caused by radiation from her dedicated work with the X-rays. She died in 1958 at only 37 years old.

James Watson and Francis Crick won a Nobel prize after Rosalind died. James Watson wrote scathing, tasteless comments about Rosalind in his book *The Double Helix*. He also admitted that he had looked at her data, and people started to figure out how the discovery really happened.

Rosalind is remembered as a woman who should have won a Nobel prize. Now that we know the story of her groundbreaking work, we can celebrate all that she accomplished!

- ALL OF THE DINING HALLS & PUBS AROUND KING'S COLLEGE WERE MEN-ONLY.
- PHOTO 51 PROVED THE DOUBLE HELIX STRUCTURE.
- LEARNED X-RAY CRYSTALLOGRAPHY IN FRANCE.
- KNEW SHE WANTED TO BE A SCIENTIST WHEN SHE WAS 15 YEARS OLD.
- CREATED A HUGE, ACCURATE TOBACCO MOSAIC VIRUS SCULPTURE FOR THE WORLD'S FAIR.
- RESEARCHED CHARCOAL TO BE USED IN GAS MASKS DURING WWII.

- Developed the RIA technique to measure hormones in the body.
- Won the 1977 Nobel Prize in Physiology or Medicine.
- Gave us a better understanding of diabetes and other hormone-related diseases.

"We must believe in ourselves or no one else will believe in us; we must match our aspirations with the competence, courage, and determination to succeed." —Rosalyn Yalow

ROSALYN YALOW
MEDICAL PHYSICIST

Rosalyn Yalow was always a fighter—her family even told stories about her standing up to her teachers when she was a child. Born in New York City in 1921, she spent her childhood going to Yankees games and reading at the library.

After completing her PhD at the University of Illinois in 1945, Rosalyn wanted to start her work in nuclear physics. The Veterans Administration Medical Center in the Bronx offered her a job figuring out ways to use radioisotopes in medicine. Without much funding, Rosalyn had to be scrappy. She turned an old janitor's closet into one of America's first radioisotope labs. Her lab partner was Solomon Berson; they would become best friends.

Rosalyn and Solomon created a new, very sensitive way to measure hormones in the body. They tagged the hormone with a radioactive isotope and then measured the amount of antibodies that were created. Their radioimmunoassay (RIA) technique is still used to learn about hormones and to screen for many different hormone-related diseases.

Rosalyn and Solomon used RIA to make new discoveries about how insulin worked inside the body, illuminating the difference between type 1 and type 2 diabetes. This helped doctors to medicate patients properly.

In 1972, Solomon died of a heart attack. Rosalyn was heartbroken; he had been like a brother to her. She understood that she would be taken less seriously now that she was a lone woman in the scientific world. Rosalyn hustled harder than ever, releasing over 60 research articles in only 4 years.

Rosalyn's hard work paid off; she was awarded many prizes and honors, including her dream, the Nobel Prize, in 1977. Her work furthered the study of endocrinology and continues to save lives to this day.

UNDERSTOOD HARD WORK AND HELPED HER MOM AT THE NECKWEAR FACTORY TO AFFORD HER OWN BRACES.

INSULIN FROM PIGS AND COWS WAS BEING USED TO TREAT DIABETES—ROSALYN FIGURED OUT WHY IT DIDN'T WORK.

RIA IS USED TO SCREEN UNBORN BABIES FOR DEADLY DISEASES, DETECT THYROID PROBLEMS, AND MAKE SURE BLOOD BANKS ARE SAFE.

TOTALLY WORTH IT!

SHE HUNG FROM THE RAFTERS TO HEAR PHYSICIST ENRICO FERMI SPEAK IN A PACKED LECTURE HALL.

RENAMED HER LAB "THE SOLOMON A. BERSON RESEARCH LAB" AFTER HIS DEATH.

KEPT CHILLED CHAMPAGNE IN HER OFFICE EVERY YEAR JUST IN CASE SHE WON THE NOBEL PRIZE.

HELPED US BETTER UNDERSTAND BACTERIA & VIRUSES.

INVENTED REPLICA PLATING TO STUDY MUTATIONS.

DISCOVERED LAMBDA PHAGE VIRUS.

PIONEER OF BACTERIAL GENETICS.

"YOU CAN BEGIN ANYTIME, EVEN THOUGH IT TAKES A LIFETIME TO BE GOOD." —ESTHER LEDERBERG

ESTHER LEDERBERG

MICROBIOLOGIST

Esther Lederberg always knew how to charm a room. Her smarts and humor made her an excellent storyteller and allowed her to get her ideas heard when they might otherwise have been ignored. She was born in 1922 in the Bronx into a very poor family. She went on to study genetics at Stanford University, where she got her master's degree in 1946. That same year she married Joshua Lederberg, a molecular biologist. Esther earned her doctorate from the University of Wisconsin, where she and Joshua would work together to study bacteria.

While peering into her microscope, Esther noticed that some of the *E. coli* bacteria cells had a "nibbled" appearance. Esther discovered a new type of bacteriophage (a virus that infects bacterium) called lambda phage. This virus acted differently; it did not immediately kill its host bacteria. Lambda phage would hide out inside the bacteria's DNA until its host was about to die; then it would spread. Studying lamda phage has given us a better understanding of RNA, DNA, and diseases like the herpes and tumor viruses.

Esther also created a new way of studying mutations in bacteria called replica plating. Before this, studying mutations took a very long time. She used a piece of velvet to stamp bacteria into new petri dishes containing different types of chemicals; it was easy to see which mutated bacteria lived or died.

This new method allowed her research team to study bacterial resistance to antibiotics and proved that bacteria can mutate spontaneously. They also found that some bacteria were resistant to antibiotics even before having contact with them. Their work led to Joshua's winning of the Nobel Prize in 1958; however, in his award speeches, he never thanked Esther for her research.

They returned to Stanford together in 1959 but divorced in 1966. She continued her work at the university and became the director of the Plasmid Reference Center. She loved her work so much that she continued her research even after she officially retired.

SHE WAS SO POOR IN GRAD SCHOOL THAT SHE ALLEGEDLY ATE THE LEFTOVER FROG LEGS FROM HER LAB DISSECTIONS.

WENT TO COLLEGE TO STUDY FRENCH LITERATURE; THEN CHANGED HER MAJOR TO BIOCHEMISTRY.

LOVED MEDIEVAL MUSIC AND FOUNDED A RECORDER ORCHESTRA.

SECOND MARRIAGE WAS TO MATTHEW SIMON, AN ENGINEER WHO ALSO LOVED MEDIEVAL MUSIC.

FIRST TRIED REPLICA PLATING WITH HER POWDER PUFF.

PUBLISHED HER DISCOVERY OF LAMBDA PHAGE IN MICROBIAL GENETICS BULLETIN IN 1951.

HELPED TO PROVE THAT BACTERIA CAN SPONTANEOUSLY MUTATE.

Statistics in STEM

The US government has used the census to understand the demographics of the American workforce. The 2011 census (published in 2013) gave the world insight into how poorly women are represented in the STEM fields. From the mid-twentieth century to the new millennium, there has been a definite increase in female scientists, but women are still underrepresented in these fields. That simply won't do. There are little girls right now who could grow up to cure cancer, explore a new galaxy, or even discover a new type of energy. Let's inspire more awesome girls and women to share their point of view and make amazing discoveries!

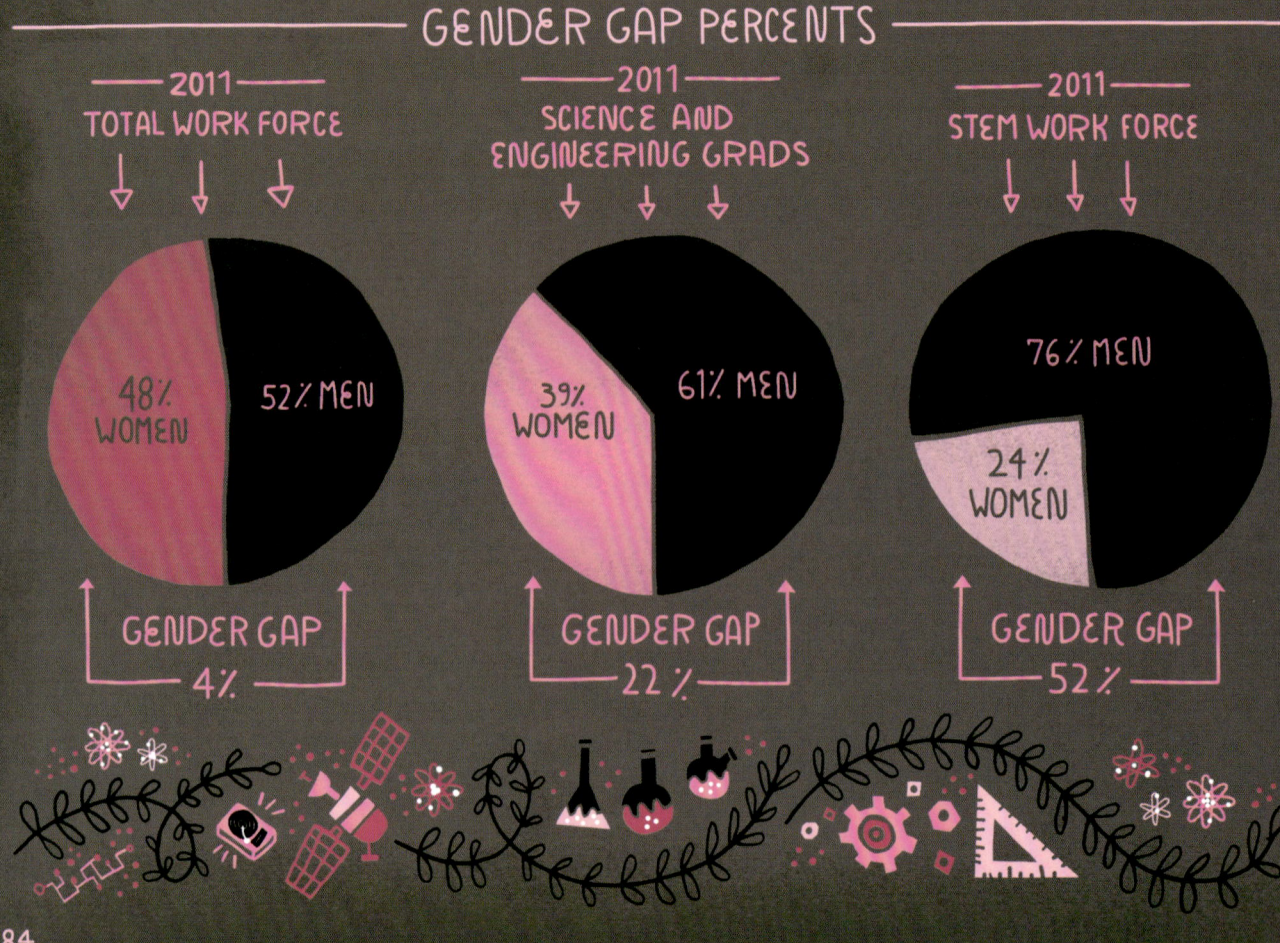

Gender Gap Percents

2011 Total Work Force
48% Women
52% Men
Gender Gap 4%

2011 Science and Engineering Grads
39% Women
61% Men
Gender Gap 22%

2011 STEM Work Force
76% Men
24% Women
Gender Gap 52%

PERCENTAGE OF WOMEN IN STEM FROM 1970-2011

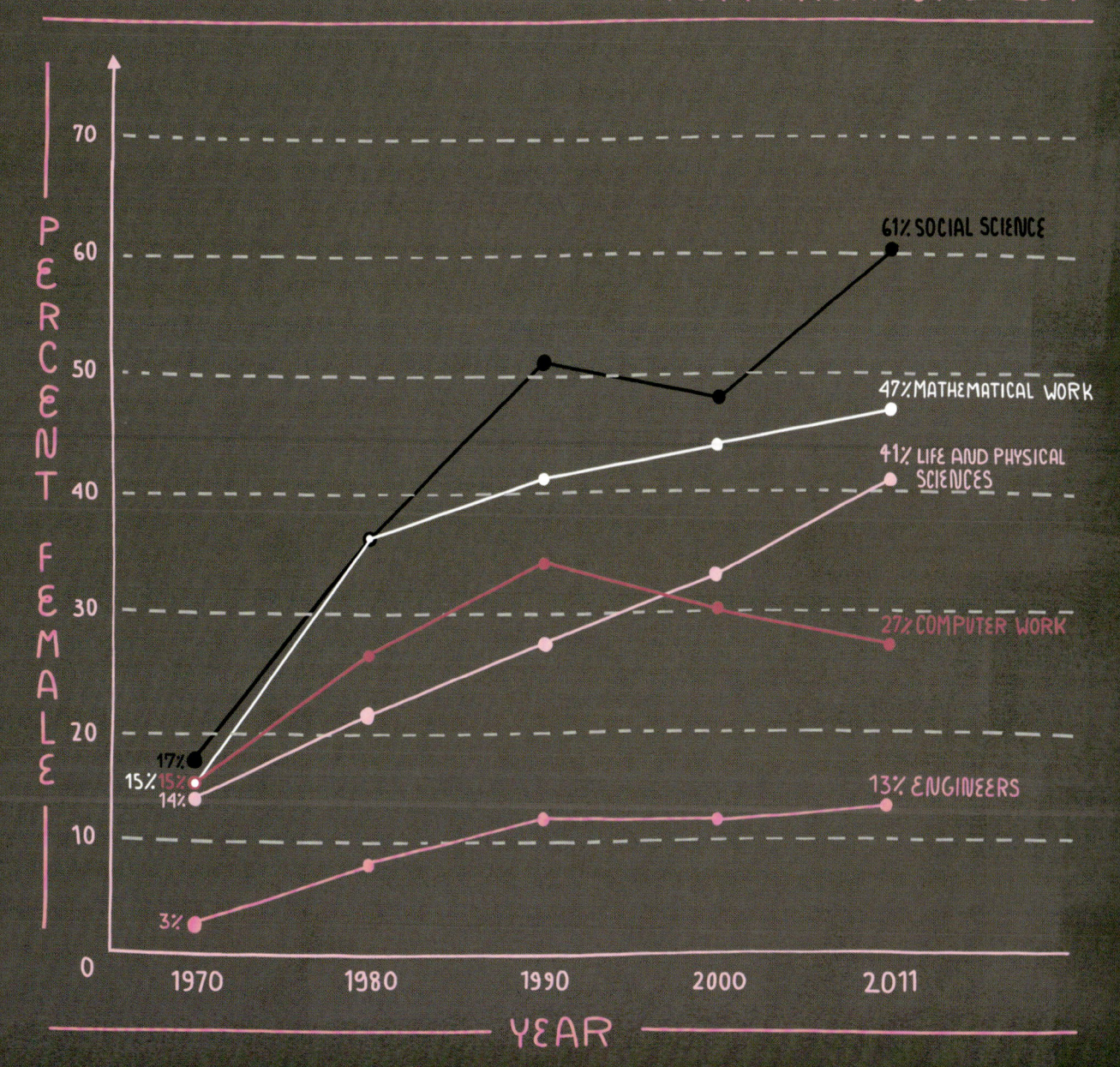

PERCENT FEMALE vs YEAR

- 61% SOCIAL SCIENCE
- 47% MATHEMATICAL WORK
- 41% LIFE AND PHYSICAL SCIENCES
- 27% COMPUTER WORK
- 13% ENGINEERS

1970 values: 17%, 15%, 15%, 14%, 3%

ELECTED TO THE NATIONAL ACADEMY OF SCIENCES.

WON THE NATIONAL MEDAL OF SCIENCE.

MADE BRAND-NEW OBSERVATIONS ON HOW GALAXIES ROTATE.

DISCOVERED REAL PROOF THAT DARK MATTER EXISTS.

— "STILL MORE MYSTERIES OF THE UNIVERSE REMAIN HIDDEN. THEIR DISCOVERY AWAITS THE ADVENTUROUS SCIENTISTS OF THE FUTURE. I LIKE IT THIS WAY." —VERA RUBIN

VERA RUBIN
ASTRONOMER

Vera Rubin was born in 1928 in Philadelphia and grew up in Washington, DC. She always had an interest in the night sky, looking up at the stars with her cardboard telescope when she was a child.

At the time she was ready to earn her master's degree, Princeton University would not admit women to their graduate astronomy program, so Vera went to Cornell University instead. At the age of 22 she made headlines and shocked scientists with her theory that the universe was rotating. Scientists today are still debating this question, though most evidence points to Vera being correct.

After earning a PhD from Georgetown University, Vera started work at Carnegie Institution of Washington, where she met Kent Ford. He invented a new spectrometer that could be used to see light from distant stars like never before and measure the Doppler effect of stars in galaxies.

Vera used his spectrometer to start her work on spiral rotating galaxies. The theory was that galaxies spin the same way solar systems do. The farther away from a gravity point, the slower an object would move, just like the different speeds of the planets circling the sun.

Vera studied over 60 different spiral galaxies. In every single one, she made the same observation: everything rotated at the same speed! What unseen form of gravity was causing this? Vera connected her findings to Fritz Zwicky's theory about undetectable "dark matter." Dark matter was creating a gravitational pull that affected how objects moved in the universe.

Although most astronomers did not believe that this invisible matter existed, Vera's findings could not be ignored. Vera's clear-cut calculations and observations could be explained only by the presence of an undetectable mass acting upon it, making her findings the strongest proof of dark matter's existence. Dark matter makes up most of the universe, but it is still a mystery to scientists today.

Vera made important observations on many galaxies throughout her career and was always willing to mentor fellow female astronomers.

HER DAD HELPED HER BUILD HER FIRST TELESCOPE.

FIRST WOMAN TO USE THE PALOMAR OBSERVATORY WITHOUT SNEAKING IN.

HAD 4 KIDS— AND ALL OF THEM BECAME SCIENTISTS.

DISCOVERED A NEW GALAXY WITH TWO HALVES THAT ROTATE IN OPPOSITE DIRECTIONS.

WON THE JAMES CRAIG WATSON MEDAL FROM THE NATIONAL ACADEMY OF SCIENCES.

SO THAT'S WHAT WE LOOK LIKE!

WISHES SHE COULD VISIT ANDROMEDA AND LOOK BACK AT THE MILKY WAY.

Did important research on alternative energy.

Helped to create software for the Centaur rocket.

Cowrote many papers about nuclear rocket engines and power plants.

——— "NOTHING WAS GIVEN TO MINORITIES OR WOMEN. IT TOOK SOME FIGHTING ———
TO GET THAT EQUAL OPPORTUNITY, AND WE'RE STILL FIGHTING TODAY." —ANNIE EASLEY

ANNIE EASLEY
COMPUTER PROGRAMMER, MATHEMATICIAN, AND ROCKET SCIENTIST

Annie Easley was born in Alabama in 1933. Living in the South at that time meant being subjected to unfair Jim Crow laws that attempted to stop African-Americans from voting. Annie used her smarts to teach others how to pass the ridiculous Jim Crow voting test. Throughout her life and career she would always give back to her community.

Annie always knew she wanted to become a nurse, and she went to school for a few years at Xavier University for a pharmacy degree. After moving to Cleveland, she planned to continue, but the pharmacy program was shut down. So she switched to mathematics—and became one of the first rocket scientists in America.

Annie heard about twin sisters who worked as human computers for the NACA (soon to become NASA). Annie knew that she could do this too and began working at NACA's Lewis Research Center in 1955. When NASA got mechanical computers, Annie used them as tools to start her work as a mathematician.

After the Russians launched Sputnik in 1957, NASA had all hands on deck working to get a rocket into space. In 1958, the Centaur project was developing a new high-energy rocket launcher. Annie worked on one of the first-ever computer programs to enable navigation in space. Since the 1960s, this upper stage of NASA rockets has been used in over a hundred launches to get satellites and probes into space. The Centaur project is still considered some of NASA's most important research.

In the 1970s, the focus of NASA went from space back to the Earth. There was an energy crisis, and scientists knew that we needed new ways to create fuel. Annie did important research on power plants and new electric batteries and created a computer program that would measure solar winds. Her work with electric batteries laid the foundation for today's hybrid vehicles.

Annie Easley understood that being flexible, believing in yourself, and working hard can lead to amazing opportunities.

I BELIEVE IN YOU!

RAISED BY A SINGLE MOM WHO ALWAYS ENCOURAGED HER.

TRAVELED TO CAPE CANAVERAL TO WATCH THE ROCKET LAUNCHES.

TUTORED UNDER-PRIVILEGED INNER-CITY KIDS IN HER FREE TIME.

WAS AN EQUAL OPPORTUNITY COUNSELOR AND TAUGHT ABOUT WORKPLACE DISCRIMINATION.

PRESIDENT OF NASA'S SKIING CLUB.

ALSO WORKED IN THE LAUNCH VEHICLES DIVISION OF NASA.

EARNED A DEGREE IN MATHEMATICS FROM CLEVELAND STATE WHILE WORKING AT NASA.

U.N. MESSENGER OF PEACE.

THE WORLD'S FOREMOST EXPERT ON CHIMPANZEES.

DISCOVERED PRIMATE TOOL MAKING.

ANIMAL RIGHTS AND WILDLIFE CONSERVATION ACTIVIST.

"ONLY WHEN OUR CLEVER BRAIN AND OUR HUMAN HEART WORK TOGETHER CAN WE REACH OUR FULL POTENTIAL." —JANE GOODALL

JANE GOODALL
PRIMATOLOGIST, ETHOLOGIST, AND ANTHROPOLOGIST

Jane Goodall was born in England in 1934. She was always curious about animals; as a young girl, she would bring earthworms into the house and she scared the chickens by trying to observe how they laid eggs.

As a young woman, Jane longed to go to Africa and study the wildlife. With no money for college, she worked as a production assistant on documentaries and as a waitress, saving up for her dream. People said that traveling to Africa was too dangerous for a woman. By pinching her pennies, Jane funded her way to Kenya. There, she met Louis Leakey, a scientist studying prehistoric humans. He was impressed with Jane's knowledge of Africa and hired her as a secretary. Louis wanted to study chimps to see if they resembled primitive man. Even though Jane was not formally educated, her unique perspective made her the best person to go to Gombe, Tanzania, to live among the chimps.

The chimps did not trust Jane. "They had never seen a white ape before," Jane said. Finally, a chimpanzee Jane named David Greybeard overcame his fear and opened up to Jane. As the chimps grew used to her, she was able to document behaviors never seen before, such as using twigs as tools. This was huge because scientists used to think that only humans used tools. Now we understand that chimps are more like us than we thought.

After Jane's famous discovery, she was sponsored by the National Geographic Society to stay in Gombe and continue observing the chimps. Through her research, she showed the world that chimps have complex social hierarchies, distinct personalities, and capacity for both compassion and cruelty. They are socially and biologically very similar to humans. Jane also knew that the chimps were in danger. Poverty had caused local communities to turn to eating chimps and destroying their habitats with bad farming practices. She started environmental conservation organizations like the Jane Goodall Institute, to help protect chimps and their habitat, and Roots & Shoots, a youth-led community action program.

Jane continues to work for world peace with the United Nations. She has changed the way we understand animals—and ourselves.

"FIGHTS TO CREATE A PROTECTED OCEAN TO STOP POLLUTION AND OVERFISHING."

IN THE JIM SUIT, SHE MADE THE DEEPEST DIVE IN 1979 & STILL HOLDS THE WOMEN'S DEPTH RECORD.

NATIONAL GEOGRAPHIC EXPLORER-IN-RESIDENCE.

HER RESEARCH, EXPLORATION, & PHOTOGRAPHY HELP EDUCATE PEOPLE ABOUT THE WORLD'S OCEANS.

"NO WATER, NO LIFE. NO BLUE, NO GREEN." – SYLVIA EARLE

SYLVIA EARLE

MARINE BIOLOGIST, EXPLORER, AND AQUANAUT

Sylvia Earle's love of the ocean has helped humankind understand it more thoroughly. Her trips to the ocean floor put her in a special group of people—like an astronaut on the moon, she has set foot on a previously unexplored frontier. She was born in 1935 in New Jersey. When she was 12 she moved to Florida, where the beaches on the Gulf of Mexico were her playground. In her quest to learn everything about the ocean, she became a marine biologist.

In 1966, Sylvia got a doctorate from Duke, where much of her research focused on the study of algae. She scuba dove to collect over 20,000 algae samples to write her dissertation. She went on to many explorations, including being the first woman to dive out of the lockout chamber of an already submerged submarine in 1968. When Sylvia was underwater, she always wanted to stay longer and explore deeper.

In 1969, a new underwater research lab called the Tektite Project was developed in which scientists could live for a few weeks—at a depth of 50 feet—in the Great Lameshur Bay in the Virgin Islands. This captured Sylvia's interest, but she could not join the all-male mission. Sylvia applied to be on the next mission and ended up leading the all-female Tektite II team the next year. She loved being able to spend up to 10 hours diving in the water among the coral reefs outside her Tektite home.

When Sylvia wasn't researching and writing books about the ocean, she was traveling around the world and exploring new depths (quite literally!). In 1979, she wore a person-sized submarine called the JIM suit and broke the depth record for an untethered dive. Deep in the Pacific Ocean off the coast of Hawaii, she observed luminescent deep sea animals. She went on to help to develop the submarine *Deep Rover* and became a National Geographic explorer-in-residence in 1998.

Throughout her career, Sylvia has focused on the fight to save our ocean. Overfishing and pollution are destroying the ocean's ecosystem and creating dead zones where no life can be sustained. Through her lectures and underwater photography, she works to ensure a protected ocean.

NICKNAMED "HER DEEPNESS" AND "THE STURGEON GENERAL."

NAMED TIME MAGAZINE'S FIRST "HERO FOR THE PLANET" IN 1998.

WITH "MISSION BLUE" SHE IS CREATING PROTECTED PARTS OF THE OCEAN CALLED "HOPE SPOTS."

SHE WAS INSPIRED BY THE BOOK HALF MILE DOWN AND JACQUES COUSTEAU'S SCUBA FILMS.

LED THE SUSTAINABLE SEAS EXPEDITIONS.

WAS THE CHIEF SCIENTIST FOR NOAA BUT LEFT SO SHE COULD HAVE MORE FREEDOM TO SPEAK OUT ABOUT OVERFISHING.

FIRST WOMAN IN SPACE.

CONTINUES TO HELP TRAIN COSMONAUTS.

LOGGED MORE TIME IN SPACE THAN ANYONE ON PREVIOUS MISSIONS.

"ONCE YOU ARE AT THIS FARAWAY DISTANCE [IN SPACE], YOU REALIZE THE SIGNIFICANCE OF WHAT IT IS THAT UNITES US. LET US WORK TOGETHER TO OVERCOME OUR DIFFERENCES." —VALENTINA TERESHKOVA

VALENTINA TERESHKOVA

ENGINEER AND COSMONAUT

Valentina Tereshkova was born in the USSR in 1937. Her family was so poor that they couldn't afford bread on their government allowance. She worked in a tire factory as a young girl and went on to work at a textile factory, but she dreamed of traveling and exploring the world.

When the Space Race between the United States and the USSR began, the USSR wanted to be first to send a woman into space. Valentina was in a parachute club, jumping out of planes for fun. She also was an enthusiastic member of the Communist Party's youth league. This made her a perfect candidate to become a cosmonaut.

Valentina was selected to compete with 4 other women. The program was so top secret that their families did not even know about it. The training was physically intense, but Valentina prevailed and was chosen to be the first woman in space.

Valentina flew solo into space on a shuttle called *Vostok VI* in 1963. She orbited the earth 48 times, setting a new record. The photographs she took in space greatly contributed to gaining a better understanding of the atmosphere.

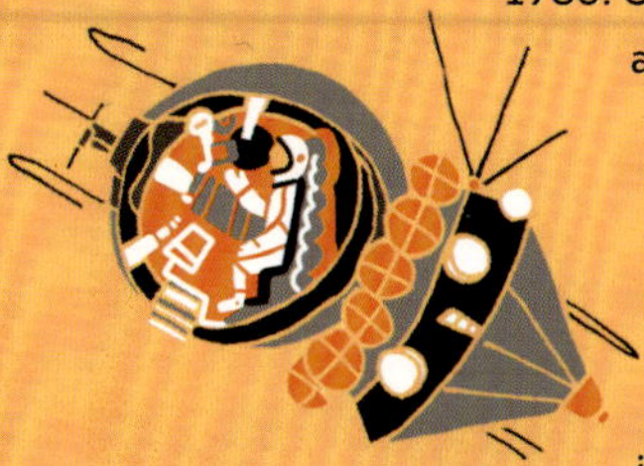

She had a bumpy ride back to Earth. There were problems in the ship's programming that she had to fix. Nauseated and disoriented, she manually corrected the error. On the way back to earth she passed out, woke up, bruised her nose, and had to stand on her head to get out of her parachute.

Valentina showed the world that women are tough as nails. After her flight, Valentina earned a doctorate in engineering and continued to work closely with aerospace engineers and the cosmonaut program. She served on the Soviet Women's Committee starting in 1968 and continues to contribute to Russian politics and work for world peace.

As a kid she wanted to travel the USSR as a train driver.

Her official call sign was "Seagull."

People brought her milk & potatoes at her landing site.

Has a lunar crater named after her.

Yelled "Hello, Sky! Take off your hat, I am on my way!" as she flew up into space.

Her first husband was cosmonaut Andriyan Nikolayev, making them the first couple who have both been in space.

Her new goal is to go to Mars.

INVENTOR OF THE LASERPHACO PROBE USED TO TREAT CATARACTS.

PIONEERED VOLUNTEER-BASED OUTREACH TO BRING EYE CARE TO IMPOVERISHED PEOPLE.

COFOUNDER OF THE AMERICAN INSTITUTE FOR THE PREVENTION OF BLINDNESS.

"BELIEVE IN THE POWER OF TRUTH... DO NOT ALLOW YOUR MIND TO BE IMPRISONED BY MAJORITY THINKING." — PATRICIA BATH

PATRICIA BATH

OPHTHALMOLOGIST AND INVENTOR

Patricia Bath was born in 1942 in Harlem, New York City. Her parents worked hard to provide her with a good education. Patricia was a genius, finishing high school in just 2 ½ years and helping with cancer research in a workshop when she was only 16. She was bound to change the world.

Patricia was no stranger to racism or sexism. She didn't know any female doctors when she was growing up, and many of the medical schools at the time were for whites only. Despite this, Patricia knew she wanted to be a doctor. After earning her medical degree from Howard University, she interned at Harlem Hospital and was accepted into Columbia University's fellowship program.

Her research showed that African-Americans were more prone to certain vision problems like glaucoma. People living in poor communities could not afford regular eye care, so relatively minor eye problems could turn into blindness. Patricia couldn't just stand by and watch this injustice, so she started the first community outreach volunteer-based eye-care program. Patricia went into parts of her hometown, Harlem, which had a high rate of poverty, and convinced a fellow surgeon to operate on patients for free. She believed that "eyesight is a human right," and she went on to cofound the American Institute for the Prevention of Blindness (AiPB).

Patricia became a professor at UCLA. She was the first female faculty member at the ophthalmology school and often did not get the respect she deserved from her peers, who assigned her an office next to where the lab animals were kept. She stood up for herself and refused that office. Eventually she became the chair of the ophthalmology residency training program, but she had had enough of dealing with the "glass ceiling" of the university. She traveled to Europe to do research and did some of her best work there.

In 1986, she finished her invention, the Laserphaco Probe, a device that removes cataracts, a major breakthrough that helped restore sight around the world. Patricia continues to work with the AiPB, bringing preventive eye care and sight-restoring surgery all around the globe.

- PATRICIA'S MOM BOUGHT HER FIRST CHEMISTRY SET.
- ORGANIZED THE FIRST MAJOR EYE SURGERY FOR HARLEM HOSPITAL IN 1970.
- VACCINATED CHILDREN IN DEVELOPING COUNTRIES AGAINST THE MEASLES.
- WAS INSPIRED TO BECOME A DOCTOR BY DR. ALBERT SCHWEITZER'S WORK WITH LEPROSY.
- VITAMIN EYE DROPS FOR BABIES
- FIRST AFRICAN-AMERICAN TO COMPLETE A RESIDENCY IN OPHTHALMOLOGY.
- FIRST AFRICAN AMERICAN WOMAN TO GET A MEDICAL PATENT IN 1988.
- COFOUNDED THE AMERICAN INSTITUTE FOR THE PREVENTION OF BLINDNESS.
- I CAN SEE! SHE RESTORED SIGHT IN PEOPLE WHO HAD BEEN BLIND FOR DECADES.

USED MUTATED FRUIT FLIES TO UNDERSTAND HOW GENES INSTRUCT STEM CELLS TO GROW.

CONTRIBUTED TO THE UNDERSTANDING OF EVOLUTION AND HOW HUMAN FETUSES DEVELOP.

WON A NOBEL PRIZE IN PHYSIOLOGY OR MEDICINE FOR HER WORK IN GENETICS.

"I IMMEDIATELY LOVED WORKING WITH FLIES. THEY FASCINATED ME, AND FOLLOWED ME AROUND IN MY DREAMS." —CHRISTIANE NÜSSLEIN-VOLHARD

CHRISTIANE NÜSSLEIN-VOLHARD
BIOLOGIST

Christiane Nüsslein-Volhard was born in Germany in 1942 and grew up in a home filled with artists, though she was more interested in studying plants and animals. She knew she wanted to be a biologist from the age of 12 and became intensely focused on pursuing her goal, even if it meant neglecting her other subjects.

In Germany at the time, men greatly outnumbered women at universities and women were expected to work in the home. It was a very competitive environment, and Christiane understood she would have to make sacrifices to put her work first. After completing her PhD in molecular biology, she decided to focus on genetics.

In her research, she worked with *Drosophila*, a.k.a. fruit flies, and was mesmerized by watching them develop. She began exploring questions about development. How does a fertilized cell become a complicated animal? How do genes instruct our stem cells to grow?

Christiane began the labor-intensive work of harvesting fly embryos and exposing them to mutagens. She then would see what part of the fly was affected by the mutation. Through tedious genetic control experiments and screenings, Christiane and her team found success: they were able to see which genes were involved in the embryo's pattern formation and which genes determined the fly's body plan and segmentation. For this work she won the 1995 Nobel Prize in physiology or medicine.

This research directly led us to understand how human embryos develop and to learn more about the evolution of species. Her work paved the way for doctors to be able to screen for birth defects and understand what can cause miscarriages.

She now uses zebra fish to research mutated genes, and she is happy to share her mutant fish with other researchers, upon request.

- DEVELOPED A BLOCK SYSTEM FOR COLLECTING FLY EMBRYOS.
- HAS AROUND 500,000 ZEBRA FISH FOR HER GENETIC RESEARCH.
- STUDIED A MUTANT FLY WITH NO HEAD AND TWO TAILS.
- LOVED GARDENING AND COLLECTED SNAILS AND BUGS AS A KID.
- NEWSPAPERS CALLED HER "LADY OF THE FLIES" AND "DAME DROSOPHILA."
- HAD DREAMS ABOUT FRUIT FLIES.
- THE CHRISTIANE NÜSSLEIN-VOLHARD FOUNDATION HELPS WOMEN SCIENTISTS PAY FOR DAYCARE.

DISCOVERED A NEW TYPE OF STAR: A PULSAR.

CONTINUES TO STUDY STARS AND BLACK HOLES.

HER RESEARCH GAVE US A FURTHER UNDERSTANDING OF THE LIFE CYCLE OF STARS AND PLANETS.

"IF WE ASSUME WE'VE ARRIVED [AT ABSOLUTE TRUTH], WE STOP SEARCHING. WE STOP DEVELOPING." —JOCELYN BELL BURNELL

JOCELYN BELL BURNELL

ASTROPHYSICIST

Jocelyn Bell Burnell was born in 1943 in Ireland. Education always came first in her home. When her secondary school wouldn't let girls into the science lab, her parents threw a fit until she was allowed in the class. Jocelyn got the best grades.

Her undergraduate studies at the University of Glasgow were challenging. She was one of very few women in the physics department. Every time she entered a science lecture, her male classmates would holler at her and make comments about her appearance. She learned to hold her head up high and hit the books. She graduated in 1965 with honors. She was accepted to the University of Cambridge's graduate program and finished her doctorate there in 1969.

At Cambridge, she joined Antony Hewish's research team and helped build a large radio telescope. She was also in charge of interpreting long, tedious printouts of radio transmissions coming from space. One night, around 2:00 a.m., she noticed a "scruff" on the readouts. It was radio waves pulsing from deep space. Her advisors thought that this could be alien life signaling from across the sky.

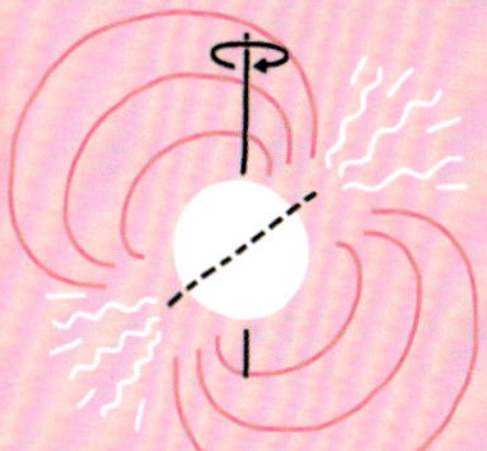

Jocelyn saw more "scruff" repeated in different places in the sky. This proved that it was not alien made, but a natural occurrence. These radio waves came from a type of small and dense star called a pulsar. This type of neutron star throws out beams of radiation like a lighthouse. Jocelyn Burnell's work helped her advisor, Hewish, to win a Nobel Prize and has been used to understand the life cycle of stars. She became one of the few female physics professors in the UK. Jocelyn still researches stars and black holes. She wants everyone to know that all elements come from exploding stars, so we "are made of star stuff."

DISCOVERED PULSARS AT THE AGE OF 24.

THE PULSAR SIGNAL WAS NICKNAMED "LGM" OR LITTLE GREEN MEN.

HAD A CHILDHOOD CAT NAMED VOSTOK, AFTER THE FIRST SATELLITES.

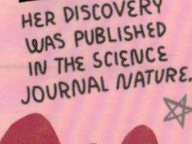

SHE WAS THE PRESIDENT OF THE ROYAL ASTRONOMICAL SOCIETY IN 2002-2004.

HER DISCOVERY WAS PUBLISHED IN THE SCIENCE JOURNAL NATURE.

SHE ADVOCATES FOR MORE WOMEN IN SCIENCE.

WAS A KEY SCIENTIST IN THE DISCOVERY OF THE HIGGS BOSON.

SHE WAS PART OF THE TEAM THAT DISCOVERED THE CHARM QUARK.

MADE IMPORTANT CONTRIBUTIONS IN THE DISCOVERY OF THE GLUON.

"I GREW UP WITH A STRONG DETERMINATION TO BE FINANCIALLY INDEPENDENT OF MEN." —SAU LAN WU

SAU LAN WU
PARTICLE PHYSICIST

Sau Lan Wu was born in the early 1940s during the Japanese occupation of Hong Kong. Although Sau Lan Wu's mother was illiterate and uneducated, she did whatever it took to make sure that Sau Lan Wu and her brother got a good education.

Against her father's wishes, Sau Lan Wu applied to 50 different colleges in America. She was accepted to Vassar College with a full scholarship in 1960—the school provided her with room, board, clothing, and books. She graduated summa cum laude and was accepted into Harvard's masters program in physics—the only woman admitted that year in her field.

After earning a PhD from Harvard, Sau Lan Wu started researching particle physics—the study of matter and how it works—at MIT, DESY, and the University of Wisconsin-Madison. Atoms are made out of protons and neutrons, which are made of quarks. Sau Lan Wu was fascinated by these particles and has dedicated her life to discovering their secrets.

With a research team led by Samuel Ting, Sau Lan Wu helped to discover the charm quark, a type of elementary particle, in 1974. After that first achievement, she became the lead on a research team that discovered the gluon, a particle that holds the quarks together.

One unanswered question in physics was how the tiny particles that make up an atom have mass. In 1964, a theory was created that mass depended on a subatomic particle named the Higgs boson—a unit of the Higgs field, which exists everywhere. The way particles interact with the field gives them more or less mass. To prove this theory, researchers faced the difficult task of finding a Higgs boson. Sau Lan Wu said, "It is like looking for a needle in a haystack—the size of a football stadium."

With a particle collider, Wu led one of the teams working to find proof of these teeny tiny subatomic particles. In 2012, her team was instrumental in observing the Higgs boson.

Sau Lan Wu is one of the most important particle physicists in her field and has made many groundbreaking discoveries. She continues to teach and research what all the stuff in the universe is made of.

The Large Hadron Collider particle accelerator is 17 miles long.

Won the European Physical Society Prize for High-Energy Physics in 1995.

Fellow of the American Academy of Arts and Sciences.

She has met her own personal goal to make at least three major discoveries.

The Higgs boson is called the "God Particle."

Summer school at the Brookhaven National Laboratory introduced her to particle physics.

Her hero is her mother.

A biography of Marie Curie inspired her to become a scientist.

WON THE 2009 NOBEL PRIZE IN PHYSIOLOGY OR MEDICINE.

FURTHERED OUR UNDERSTANDING OF HUMAN LIFE SPAN AND CHROMOSOMES.

DISCOVERED TELOMERASE, THE ENZYME THAT REBUILDS TELOMERES.

"DON'T BE AFRAID TO ASK PEOPLE FOR HELP — AND THEN FEEL FREE TO IGNORE IT!" — ELIZABETH BLACKBURN

ELIZABETH BLACKBURN
MOLECULAR BIOLOGIST

Elizabeth Blackburn was born in 1948 in Tasmania, Australia. She played with any animal she could get her hands on—tadpoles, jellyfish, rabbits, and chickens all became her playmates. Her love of animals led to her passion for biology.

After Elizabeth completed her master's degree in Australia, she left her home to earn a PhD in the UK. At the University of Cambridge, she studied DNA sequences of bacteriophages for her dissertation. She was thrilled to be working with DNA, realizing it was the key to understanding how all life works. She went to America to continue pursuing research in her new favorite subject.

In the 1970s, no one really knew what the ends of chromosomes were like—under the microscope, they just seemed like blurry blobs. Chromosomes are extremely important and exist in each of our cells. They are tightly wound DNA material that tells our cells what they are supposed to do in our body. Elizabeth wanted to fully understand how they worked.

Elizabeth noticed that there was a special kind of DNA called telomeres on each end of the chromosomes that worked as a protective cap. She discovered that telomeres are made of nonessential repeating segments of DNA that break off a little bit every time a cell divides, protecting the important information. When we get older, this protective cap wears out and our chromosomes become damaged. This loss of DNA information causes our cells to not work correctly or die, leading to diseases like cancer, organ failure, and Alzheimer's.

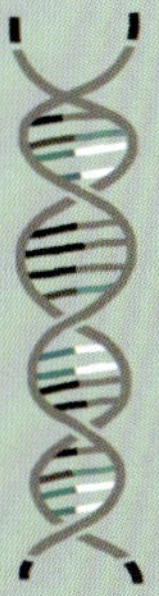

Elizabeth wanted to understand what keeps our bodies' telomeres healthy. In 1984, with the help of her grad student Carol Greider, she codiscovered telomerase, an enzyme that rebuilds telomeres to a healthy length. In 2009 Elizabeth was awarded the Nobel Prize in physiology or medicine.

Elizabeth Blackburn's research shows that keeping a healthy telomere length is directly responsible for living a long, healthy life. It is not a magical solution, though; too much telomerase leads to cancer, and too little causes the effects of old age. Elizabeth described it as "living on a knife's edge." She continues to study telomerase and telomeres, working to figure out the science behind longevity.

EXERCISE, SLEEP, LOW STRESS LEVELS, AND HEALTHY DIET ARE PROVEN TO HELP KEEP TELOMERES HEALTHY.

WORKED AT YALE, UC SAN FRANCISCO, AND UC BERKELEY.

I'M POND SCUM!

WORKED WITH A PROTOZOAN CALLED TETRAHYMENA TO STUDY TELOMERES.

WAS THE PRESIDENT OF THE AMERICAN SOCIETY FOR CELL BIOLOGY IN 1998.

WAS IN THE 2007 PUBLICATION OF "THE TIME 100—THE PEOPLE WHO SHAPED OUR WORLD" IN TIME MAGAZINE.

ELIZABETH MET BARBARA MCCLINTOCK, WHO TOLD HER TO TRUST HER OWN INTUITION!

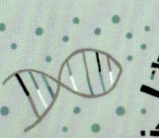

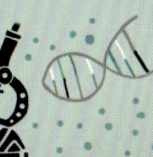

PIONEERED VOLCANO NATURE PHOTOGRAPHY.

STARTED HER OWN FOUNDATION FOR VOLCANOLOGY WITH HUSBAND, MAURICE KRAFFT.

USED OBSERVATIONS TO HELP GOVERNMENTS DEVELOP VOLCANO EVACUATION PROCEDURES.

"FOR ME THE DANGER IS NOT IMPORTANT...ON VOLCANOES I FORGET EVERYTHING." —KATIA KRAFFT

KATIA KRAFFT
GEOLOGIST AND VOLCANOLOGIST

Katia Krafft was born in 1942 in France. She fell in love with volcanoes when she saw pictures of them. She studied geology at the University of Strasbourg, where she also met her husband and fellow volcano fanatic, Maurice Krafft.

Katia started her career by taking gas samples of volcanoes, and she and Maurice would document volcanoes erupting by observing them in person. Volcanoes are unpredictable and dangerous, and many scientists were too afraid to observe eruptions in person, but not Maurice and Katia. Throughout the 1970s and 1980s, they documented the volcanoes. Katia would photograph them while Maurice captured them on video.

Katia and Maurice's observations have led to a better understanding of volcanic eruptions. They took viscosity measurements and gas readings and collected mineral samples just feet away from erupting volcanoes. They documented how these eruptions affected the ecosystems. Together they witnessed and documented new volcanoes being formed, the effects of acid rain, and dangerous ash clouds. They even went on a raft into a lake of acid to get proper readings. Their photography and videos allowed them to work with local governments on safety procedures and evacuations. Some of their last videos were *Understanding Volcanic Hazards* and *Reducing Volcanic Risks*, but that didn't mean they stopped being daredevils. They continued to push the boundaries to get their observations, going closer to the volcano and staying longer during an eruption. In 1991, their luck ran out, and the volcano Mount Unzen in Japan took Katia and Maurice's lives, along with those of 41 other scientists and journalists, when the pyroclastic flow changed course.

Katia died doing what she loved, with the person she loved. For years she studied volcanoes right at their edge. Her bravery and expertise have given us a greater understanding of volcanoes that will endure.

THE KRAFFT MEDAL IS NOW GIVEN OUT TO EXCEPTIONAL VOLCANOLOGISTS.

WORE A SPECIAL HELMET TO PROTECT HER SKULL FROM FALLING ROCKS.

MADE A DOCUMENTARY, THE VOLCANO WATCHERS, FOR THE PBS SHOW NATURE.

KATIA AND MAURICE STARTED THEIR OWN VOLCANO CENTER IN 1968.

TOGETHER THE KRAFFTS WROTE MANY BOOKS THAT FUNDED THEIR TRIPS ALL OVER THE WORLD.

WAS KILLED BY A PYROCLASTIC FLOW THAT CHANGED DIRECTION.

FIRST AFRICAN-AMERICAN WOMAN IN SPACE.

PRINCIPAL OF THE 100 YEAR STARSHIP PROJECT.

FOUNDER OF JEMISON GROUP INC. AND BIOSENTIENT CORPORATION.

"THE FIRST THING ABOUT EMPOWERMENT IS TO UNDERSTAND THAT YOU HAVE THE RIGHT TO BE INVOLVED. THE SECOND ONE IS THAT YOU HAVE SOMETHING IMPORTANT TO CONTRIBUTE. AND THE THIRD PIECE IS THAT YOU HAVE TO TAKE THE RISK TO CONTRIBUTE IT." — MAE JEMISON

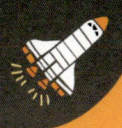

MAE JEMISON

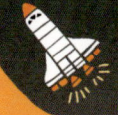

ASTRONAUT, EDUCATOR, AND DOCTOR

Mae Jemison always knew she would go into space. She was born in 1956 in Alabama and grew up in Chicago. She was obsessed with the Apollo missions but noticed that there was no one who looked like her going up into space. However, the fictional TV show *Star Trek* featured people of different genders and races working together. This had an impact on young Mae, and Lieutenant Uhura became her role model.

Mae went to Stanford and double majored in chemical engineering and African-American studies. She went on to Cornell and became a medical doctor. She worked in the Peace Corps in Sierra Leone and Liberia for several years. She continued working as a doctor until it was time to chase her space dream. Mae applied to NASA and became an astronaut.

In 1992, Mae Jemison became the first African-American woman in space. On the space shuttle *Endeavour*, she took an Alpha Kappa Alpha sorority flag, a West African Bundu statue, and a poster of Judith Jamison dancing. She wanted African and African-American culture to be represented in space and no longer left out.

The following year, she left NASA and started numerous companies, including her own technology consulting firm, the Jemison Group Inc. Mae is the founder of the BioSentient Corporation, which creates devices that will allow doctors to monitor patients' day-to-day nervous system functions.

The technology and problem solving to get humans in space created inventions that we use today on earth. Mae was inspired by this and became principal of the 100 Year Starship project. The goal is to make sure human beings will be able to travel to the next solar system within the next 100 years. This project will also inspire new solutions to materials, recycling, energy, and fuel, just as the space race did. Dr. Mae Jemison still has her eyes on the stars while helping solve problems here on earth.

FOUND OUT SHE WAS GOING TO BE AN ASTRONAUT IN BETWEEN GIVING MEDICAL EXAMINATIONS.

HER DAD TAUGHT HER HOW TO COUNT CARDS AS A KID.

WENT ON AN EIGHT-DAY MISSION IN SPACE.

THE FIRST LANDMARK SHE IDENTIFIED FROM SPACE WAS CHICAGO, HER HOMETOWN.

DID EXPERIMENTS WITH BONE CELLS WHILE IN SPACE.

WON A SCHOLARSHIP TO STANFORD WHEN SHE WAS 16.

WAS FEATURED ON AN EPISODE OF STAR TREK: THE NEXT GENERATION.

FOUNDED "THE EARTH WE SHARE" SCIENCE CAMP FOR KIDS.

SHE IS A DANCER.

COFOUNDER OF KAVLI INSTITUTE

DISCOVERED GRID CELLS AND HOW MAPS ARE MADE IN THE HUMAN MIND

WON A NOBEL PRIZE IN PHYSIOLOGY OR MEDICINE WITH HER HUSBAND EDVARD.

"A GOOD DESIGNER HAS A LOT IN COMMON WITH A GOOD RESEARCHER. BOTH HUNT FOR EXCELLENCE AND PERFECTION. AND YOU HAVE TO REALLY FOCUS ON THE DETAILS, AND YOU DON'T REALLY KNOW WHAT THE FINAL RESULT WILL BE BEFORE YOU HAVE IT." —MAY-BRITT MOSER

MAY-BRITT MOSER

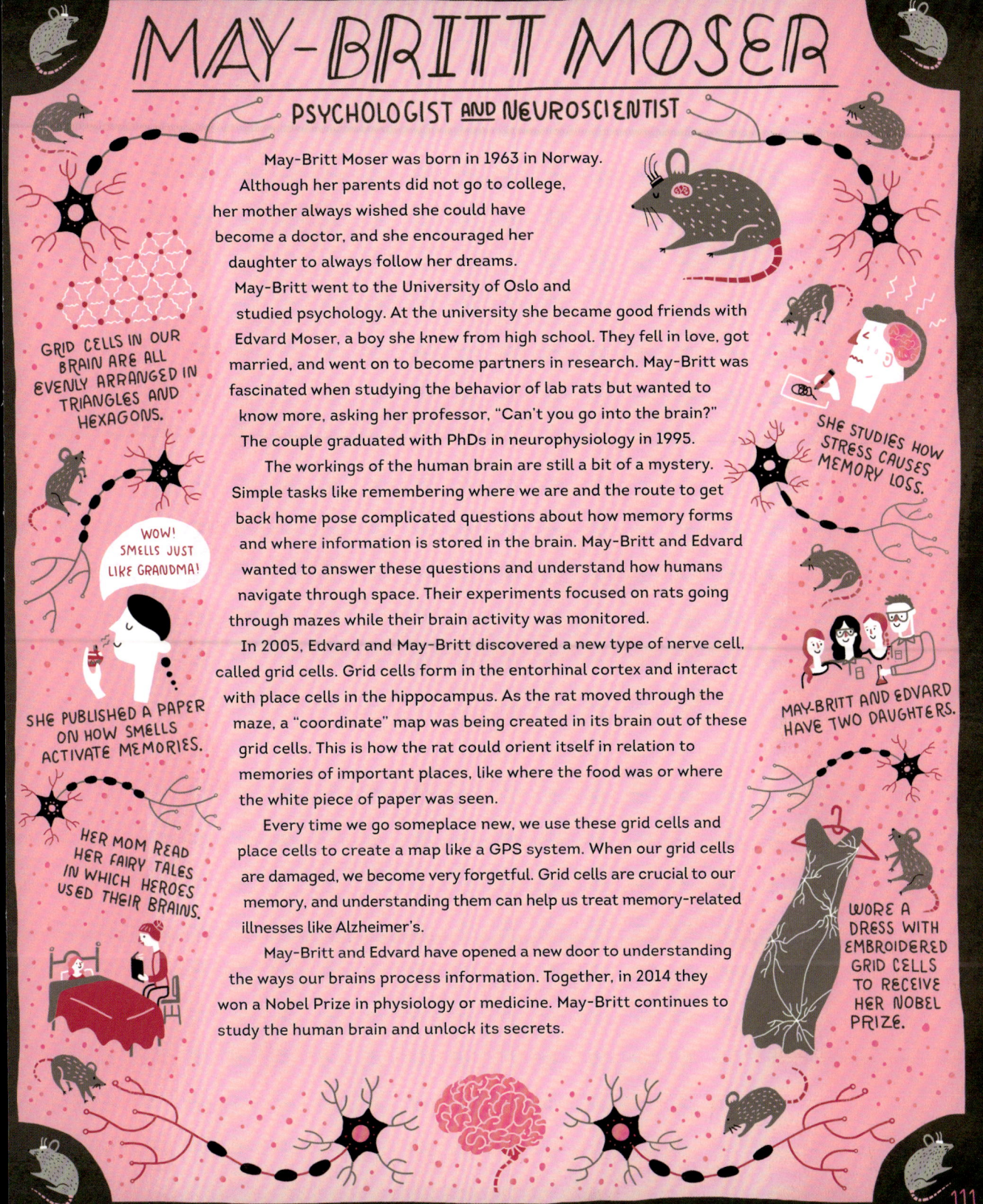

PSYCHOLOGIST AND NEUROSCIENTIST

May-Britt Moser was born in 1963 in Norway. Although her parents did not go to college, her mother always wished she could have become a doctor, and she encouraged her daughter to always follow her dreams.

May-Britt went to the University of Oslo and studied psychology. At the university she became good friends with Edvard Moser, a boy she knew from high school. They fell in love, got married, and went on to become partners in research. May-Britt was fascinated when studying the behavior of lab rats but wanted to know more, asking her professor, "Can't you go into the brain?" The couple graduated with PhDs in neurophysiology in 1995.

The workings of the human brain are still a bit of a mystery. Simple tasks like remembering where we are and the route to get back home pose complicated questions about how memory forms and where information is stored in the brain. May-Britt and Edvard wanted to answer these questions and understand how humans navigate through space. Their experiments focused on rats going through mazes while their brain activity was monitored.

In 2005, Edvard and May-Britt discovered a new type of nerve cell, called grid cells. Grid cells form in the entorhinal cortex and interact with place cells in the hippocampus. As the rat moved through the maze, a "coordinate" map was being created in its brain out of these grid cells. This is how the rat could orient itself in relation to memories of important places, like where the food was or where the white piece of paper was seen.

Every time we go someplace new, we use these grid cells and place cells to create a map like a GPS system. When our grid cells are damaged, we become very forgetful. Grid cells are crucial to our memory, and understanding them can help us treat memory-related illnesses like Alzheimer's.

May-Britt and Edvard have opened a new door to understanding the ways our brains process information. Together, in 2014 they won a Nobel Prize in physiology or medicine. May-Britt continues to study the human brain and unlock its secrets.

- GRID CELLS IN OUR BRAIN ARE ALL EVENLY ARRANGED IN TRIANGLES AND HEXAGONS.
- WOW! SMELLS JUST LIKE GRANDMA!
- SHE PUBLISHED A PAPER ON HOW SMELLS ACTIVATE MEMORIES.
- HER MOM READ HER FAIRY TALES IN WHICH HEROES USED THEIR BRAINS.
- SHE STUDIES HOW STRESS CAUSES MEMORY LOSS.
- MAY-BRITT AND EDVARD HAVE TWO DAUGHTERS.
- WORE A DRESS WITH EMBROIDERED GRID CELLS TO RECEIVE HER NOBEL PRIZE.

HAS DONE IMPORTANT WORK IN HYPERBOLIC GEOMETRY.

FIRST WOMAN TO WIN THE FIELDS MEDAL

HAS GIVEN US NEW INSIGHT INTO THE DYNAMICS OF ABSTRACT SURFACES.

"YOU HAVE TO SPEND SOME ENERGY AND EFFORT TO SEE THE BEAUTY OF MATH." —MARYAM MIRZAKHANI

MARYAM MIRZAKHANI
MATHEMATICIAN

Maryam Mirzakhani, born in 1977 in Iran, grew up reading every book she could find. She wanted to become a writer and didn't have much of an interest in math until high school, when she got her hands on the entrance questionnaire for an international math competition. Maryam struggled to solve the problems and spent days on a worksheet that should have taken her hours. Excited by this new challenge, she demanded that her all-girl high school provide the same math courses as the boys' schools did.

Maryam came to America for graduate school at Harvard. She became interested in understanding the surface of a shape and what happens when it is distorted. She enjoyed finding the beauty in mathematics and focused on hyperbolic surfaces.

Hyperbolic doughnuts are abstract shapes; to understand them, you need to find straight lines, or "simple" geodesics, inside. This is incredibly difficult. Maryam created an equation that showed the relationship between the amount of simple geodesics and the length of the side of a hyperbolic structure. Her work is fundamental in understanding curved shapes and surfaces.

There was another unsolved problem in mathematics: a billiard ball is bouncing around, hitting the sides of a table forever in a frictionless environment. Will a ball that is hit from any direction always end up where it started? What about the infinite possible shapes of the billiard table? This problem was so complicated, computers couldn't even simulate it!

Maryam thought of a different way to solve this problem. Instead of moving the ball around the table, she mirrored the table around the ball. When the ball hit a side, the table would flip and change angles, so it would look as if the ball stayed in a straight line. She figured out that the ball will always close its loop. This has been compared to how particles might behave and has given us a better understanding of geometry, physics, and quantum theory.

In 2014, Maryam won the Fields Medal for her work, the first woman so honored. Maryam works at Stanford, where she continues to push boundaries in mathematics.

THE FIELDS MEDAL IS CONSIDERED THE NOBEL PRIZE IN MATH.

HAS DONE IMPORTANT WORK ON TEICHMÜLLER DYNAMICS AND MODULI SPACE.

MARYAM AND HER FRIEND BECAME THE FIRST GIRLS ON IRAN'S INTERNATIONAL MATHEMATICAL OLYMPIAD TEAM AND SHE WON A GOLD MEDAL.

AS A CHILD SHE WAS INSPIRED WHEN HER OLDER BROTHER TOLD HER ABOUT THE MATH PROBLEM OF ADDING ALL THE NUMBERS BETWEEN 1 AND 100.

$\frac{N(N+1)}{2}$

CREATED A NEW PROOF OF EDWARD WITTEN'S THEORY ABOUT TOPOLOGICAL MEASUREMENTS OF MODULI SPACES.

SHE DRAWS HYPERBOLIC SHAPES ON HUGE PIECES OF PAPER TO BETTER UNDERSTAND THEM.

MORE WOMEN IN SCIENCE

IRÈNE JOLIOT-CURIE
1897-1956
Daughter of Marie Curie and also a Nobel Prize winner in chemistry. Discovered a way to create synthetic radioactive elements in the lab.

JANAKI AMMAL
1897-1984
Botanist who did important work to crossbreed sugar cane and worked at the Botanical Survey of India.

ANNA JANE HARRISON
1912-1998
Studied how atoms become molecules and was the first woman president of the American Chemical Society.

SHIRLEY ANN JACKSON
1946-
Physicist who is president of Rensselaer Polytechnic Institute and the first African-American to earn a doctorate from MIT.

LINDA BUCK
1947-
Won the Nobel prize in physiology or medicine for her work on how we use our olfactory nerves to understand smells.

FRANÇOISE BARRE-SINOUSSI
1947-
Virologist who won the Nobel Prize in physiology or medicine for her discovery of HIV.

MARIA MITCHELL
1818-1889
First American woman to work as an astronomer. Discovered "Miss Mitchell's Comet."

EMILY ROEBLING
1843-1903
American field engineer responsible for the execution of the Brooklyn Bridge.

SOFIA KOVALEVSKAYA
1850-1891
Russian mathematician who worked on partial differential equations and created the Cauchy-Kovalevskaya theorem.

MARY LEAKEY
1913-1996
Her fossil discoveries of our ancient ancestors, or "missing links," changed our understanding of human evolution.

EDITH FLANIGEN
1929-
Chemist who invented ways to process crude oil and purify water using molecular sieves, and ways to grow new materials like synthetic emeralds.

ADA YONATH
1939-
Israeli crystallographer who discovered the structure of ribosomes and won the Nobel prize in chemistry in 2009.

SALLY RIDE
1951-2012
First American woman in space and director of the California Space Institute.

TESSY THOMAS
1963-
Indian engineer who was instrumental in creating the most powerful long-range nuclear missile ever.

THE NEXT GREAT SCIENTIST COULD BE YOU!
Women everywhere are working hard, learning, and researching to make the next big breakthrough.

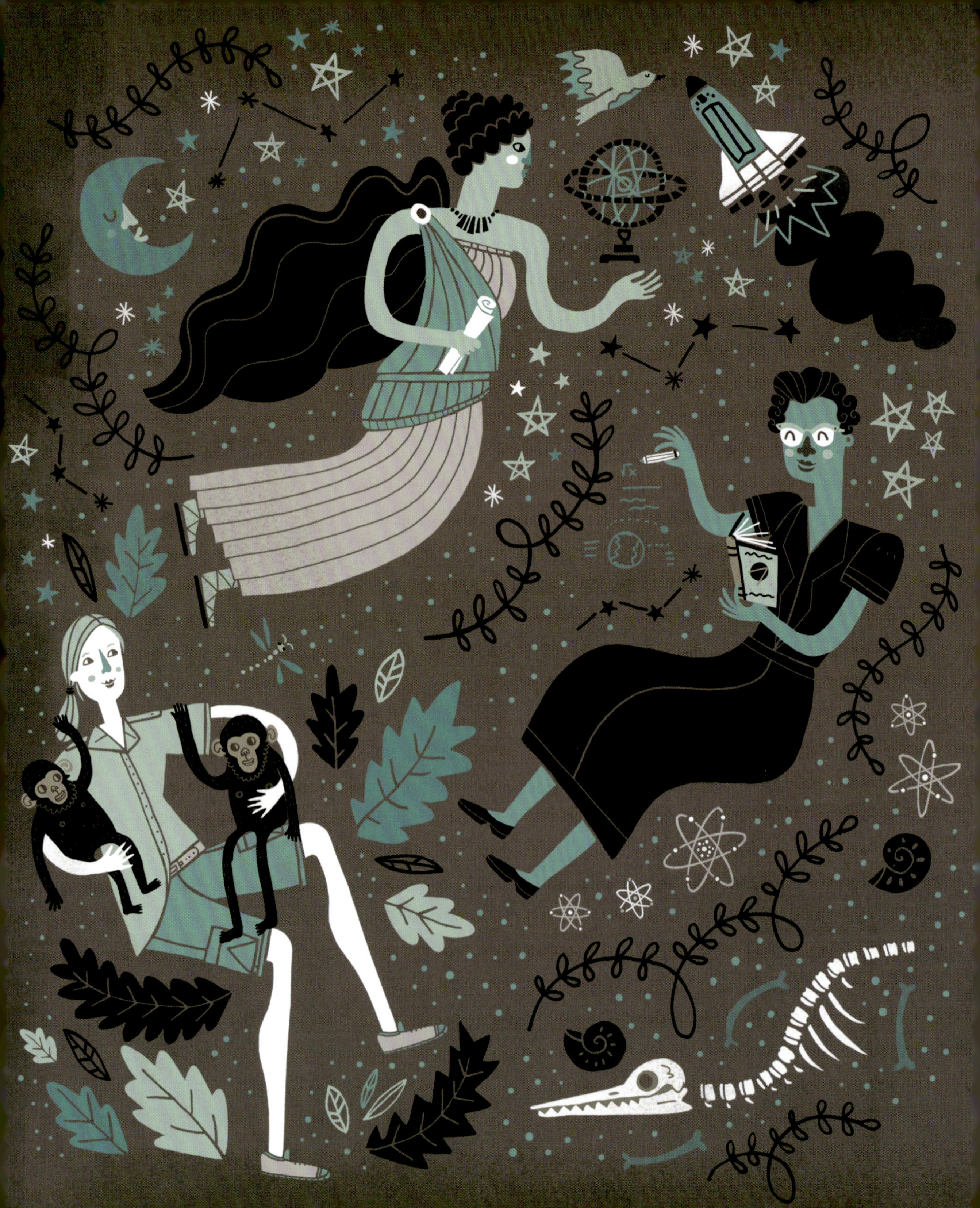

CONCLUSION

Women make up half of our population, and we simply cannot afford to ignore that brain power—the progress of humankind depends on our continual search for knowledge. The women in this book prove to the world that no matter your gender, your race, or your background, anyone can achieve great things. Their legacy lives on. Today, women all over the world are still risking everything to discover and explore.

Let us celebrate these trailblazers so we can inspire the next generation. Together, we can pick up where they left off and continue the search for knowledge.

So go out and tackle new problems, find your answers, and learn everything you can to make your own discoveries!

GLOSSARY

ABOLITIONIST
An activist working to end slavery and the slave trade.

ANTIVIRALS
Drugs to specifically fight viral infections.

ATOM
The smallest unit of matter. The center, or nucleus, is made out of positive protons and neutral neutrons. The nucleus is surrounded by negatively charged electrons flying around. When different kinds of atoms combine, they make molecules.

BACTERIA
A type of single-celled organism found everywhere. There are many different kinds and they can be useful, harmful, or helpful to plants and animals. For example, some make us sick, some help us digest our food, and some help turn milk into cheese.

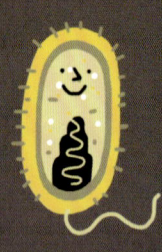

BACTERIOPHAGE
A virus that attacks and infects bacteria and then reproduces inside of it.

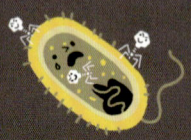

BETA DECAY
A type of radioactive decay of an atom in which a proton changes into a neutron (or vice versa) and a beta particle is emitted.

BOTANY
The study of plants.

CELL
The smallest unit of life. It can live on its own as an amoeba or bacteria. Cells are also the building blocks for tissues to create organs in plants and animals.

CHROMOSOME
Tightly wound strands of DNA bundled together. They are in the nucleus of cells, instructing the cells how to work.

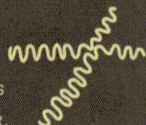

COMPILER
A computer program that translates a computer language, like COBOL, into something machines can understand.

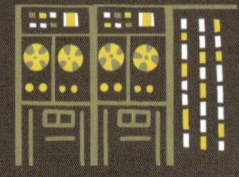

DNA
This molecular strand is our genetic instructions. It is inherited from our parents and tells our cells and bodies how to grow, reproduce, and function. All organisms have DNA, and the strands are found in the nucleus of each cell.

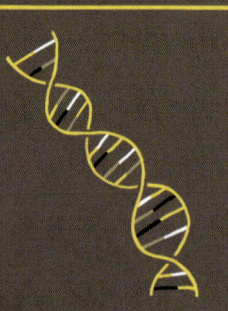

ECLIPSE
A phenomenon that happens when 3 objects in space line up and the one in the middle blocks the view or light of one outer object from reaching the other. For example, in a lunar eclipse, the Earth aligns between moon and sun, casting its shadow onto the moon and blocking the sun's light; in a solar eclipse, the moon aligns between Earth and sun, casting its shadow onto the Earth and blocking the view of the sun and its light from the Earth.

ECOSYSTEM
A group of organisms living together and their interaction with each other and the air, water, and soil around them.

ELECTRIC ARCS
When 2 electrical currents ionize the gas or air around or between them, it creates a plasma discharge. They now can move through the air, which normally does not conduct electricity. Lightning is an example of a naturally occurring electric arc.

ELEMENT
In chemistry, a substance made up of only one kind of atom—for example, gold or helium.

ERGONOMICS
The study of how people interact with tools and their environment. Ergonomics helps in the design of tools that work comfortably with how our bodies move.

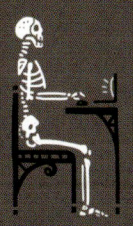

FOSSIL
The remains of ancient animals and plants that have been preserved or even petrified over time. Sometimes the fossil, like an old dinosaur bone, is stuck in a rock. Sometimes it is an imprint in the rock, like a footprint.

FREUDIAN THEORY
Part of a branch of social science called psychiatry. Named after the father of modern psychoanalysis, Sigmund Freud, it is a theory of how our unconscious desires interact with our consciously chosen actions.

GENETICS
The study of how our DNA, chromosomes, and genes work, how the genes that are passed down from our ancestors and parents change over time, and how they affect organisms.

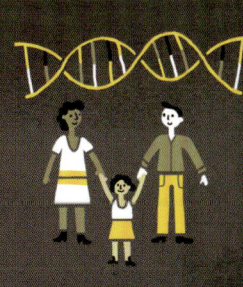

GEOMORPHOLOGY
The study of how the surface of the Earth has changed over the span of its existence: for example, how mountains and continents form.

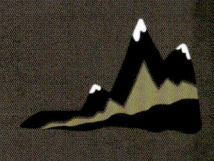

GLUCOSE
A sugar molecule that is an important source of energy for people. For example, when you eat a doughnut, all the sugars and carbohydrates are digested and broken down into glucose.

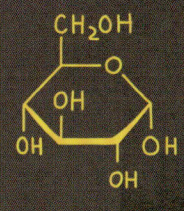

HERPETOLOGY
The study of reptiles and amphibians.

HUMAN COMPUTER
Before we had mechanical computers, complicated math equations were done by a large group of people. Each person got a small part of the equation, and together they could solve the problem.

ISOTOPES
Created when the amount of neutrons in an atomic nucleus changes. There can be many different isotopes of the same atom, all with a different atomic mass but with the same number of protons.

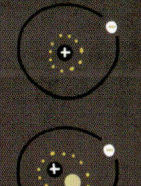

INSULIN
The hormone that lets our body process sugar, or glucose, for energy and storage.

KOMODO DRAGON
The largest species of lizard, which can be very dangerous and venomous. It is native to Indonesia.

LACTIC ACID
A molecule created in our muscles when we exercise. It is created during the Cori cycle, described by Gerty and Carl Cori.

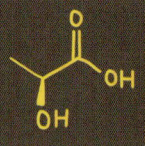

THE MANHATTAN PROJECT
A top-secret project created by the United States during World War II to develop the atomic bomb.

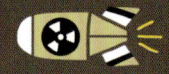

METAMORPHOSIS
Process in which an animal changes dramatically from one life stage to another—for example, a caterpillar using a cocoon to become a butterfly.

MODULI SPACE
Some math problems have more than one answer. The set of all the possible answers to a particular geometry problem is called the moduli space.

MUTATIONS
A permanent change in the sequence of genes in an organism. It can happen while a cell is dividing its DNA during reproduction when parts can be deleted or added to the code.

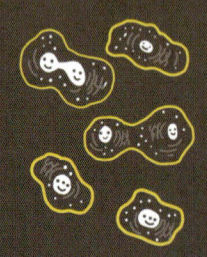

NASA
The US National Aeronautics and Space Administration.

NERVE CELLS
Also known as neurons, these are the cells that send information to our brain through chemical and electrical signals. These cells allow us to feel sensations and have memories and thoughts, and they tell our body to move.

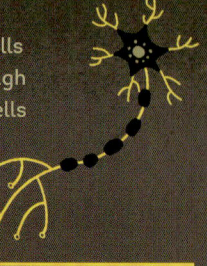

NERVE GROWTH FACTOR
A protein important for growing new cells and repairing and maintaining our nerve cells. It circulates throughout our entire body and is important for our survival.

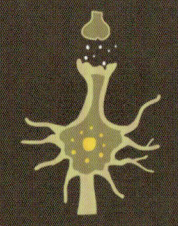

NOBEL PRIZE
An annual prize in the subjects of physics, chemistry, physiology or medicine, literature, economics, and peace. It is seen globally as one of the most honorable prizes to be won.

NOETHER THEOREM
Proved that whenever there is a physical action that involves a predictable symmetry, it is because there is a law of conservation (for example, of mass, energy, momentum, etc.).

PARTICLE ACCELERATOR
Uses an electromagnetic field to make particles move at super-fast speeds and smash apart when they collide with each other.

PULSARS
A neutron star that emits a beam of electromagnetic radiation. The beams come out of the magnetic poles of the star, and, as the star rotates, the beam pulses like a lighthouse.

PUNCH CARDS
Literally a piece of stiff paper with holes punched into it in different places, creating code. It was one of the first methods for talking to a machine or a computer.

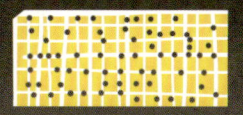

QUARKS
A type of subatomic particle that makes composite particles. In fact, they create neutrons and protons. Today we know 6 kinds of quarks, called flavors: up, down, strange, charm, bottom, and top. There is still a lot to be discovered about quarks.

RADIOACTIVITY
The energy released when the atomic nucleus changes. This release can include alpha particles, beta particles, gamma rays, and electromagnetic waves.

RADIATION TECHNOLOGY
Radiation technology is used to see broken bones and in cancer treatment, but too much radiation exposure can cause cancer or radiation poisoning.

RING THEORY
The study of "rings." In math, rings are sets of numbers where addition and multiplication are defined.

SOCIAL HIERARCHY
How animals or human organize themselves to establish dominance and access to food and resources.

SPECTROSCOPE
An instrument that uses a prism to break light into the rainbow of colors across the electromagnetic spectrum. It is used in astronomy and chemistry because atoms absorb light at different frequencies. By breaking up the light, measuring the different intensity and wavelengths, and looking for black line breaks, a scientist can pick out different atoms in the light.

STELLAR SPECTRA
The rainbow of light and dark breaks seen from a star when looking through a spectroscope.

SUFFRAGIST
An activist who fought for voting rights for women.

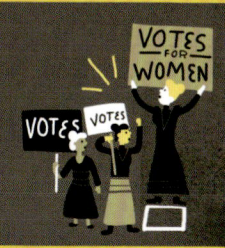

VIRUS
Infectious agent, smaller than a cell and not considered living. It can reproduce only by infecting other cells and in doing so causes diseases.

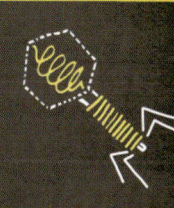

X-RAY CRYSTALLOGRAPHY
A tool that uses an X-ray beam on a crystal version of a substance. The beam goes in all different directions. By measuring the angles of the beams, scientists can understand the 3-dimensional structures of different molecules and atoms.

SOURCES

Researching this book was so much fun. I used all sorts of sources: newspapers, interviews, lectures, books, films, and the Internet! If you are interested in learning more about these women (and you should!), here are some of the sources I consulted. For more resources on the specific women featured in this book, go to www.rachelignotofskydesign.com/women-in-science-resources.

FILMS

Beautiful Minds: Jocelyn Bell Burnell. Directed by Jacqui Farnham. BBC Four, 2010. Series 1, episode 1 of 3.

Commencement Address: From Vassar to the Discovery of the Higgs Particle. Performed by Sau Lan Wu. Vassar College, 2014. commencement.vassar.edu/ceremony/2014/address/.

The Genius of Marie Curie. Directed by Gideon Bradshaw. BBC, 2013.

Great Floridians Film Series–Marjory Stoneman Douglas. By Marilyn Russell. Florida Department of State, 1987.

Jane Goodall at Concordia: Sowing the Seeds of Hope. Concordia University, 2014. www.youtube.com/watch?v=vibssrQKm60.

May-Britt and Edvard Moser–Winner of the Körber European Science Prize 2014. Directed by Axel Wagner. Koerber-Stiftung, 2014. www.youtube.com/watch?v=592ebE5U7c8.

Mission Blue. Directed by Robert Nixon and Fisher Stevens. Insurgent Media, 2014.

Signals: The Queen of Code. Directed by Gillian Jacobs. FiveThirtyEight, 2015. fivethirtyeight.com/features/the-queen-of-code/.

Valentina Tereshkova: Seagull in Space. Russia Today, 2013. www.youtube.com/watch?v=Y2k9s-NbNaA.

The Volcano Watchers. Directed by David Heeley. PBS, 1987.

WEBSITES

American Museum of Natural History: www.amnh.org
Encyclopedia Britannica: www.britannica.com
Jewish Women's Archive: www.jwa.org/encyclopedia
MAKERS, The largest video collection of women's stories: www.makers.com
NASA: www.nasa.gov
National Inventors Hall of Fame: www.invent.org
National Women's History Museum: www.nwhm.org
The Official Website of the Nobel Prize: www.nobelprize.org
Psychology's Feminist Voices: www.feministvoices.com
US National Library of Medicine: www.nlm.nih.gov/changingthefaceofmedicine

BOOKS

Adams, Katherine H., and Michael L. Keene. 2010. *After the Vote Was Won: The Later Achievements of Fifteen Suffragists*. Jefferson, NC: McFarland.

Layne, Margaret. 2009. *Women in Engineering*. Reston, VA: ASCE Press.

Dzielska, Maria. 1995. *Hypatia of Alexandria*. Cambridge, MA: Harvard University Press.

McGrayne, Sharon Bertsch. 1993. *Nobel Prize Women in Science: Their Lives, Struggles, and Momentous Discoveries*. Secaucus, NJ: Carol Publishing Group.

Peterson, Barbara Bennett. 2000. *Notable Women of China: Shang Dynasty to the Early Twentieth Century*. Armonk, NY: M.E. Sharpe.

Swaby, Rachel. 2015. *Headstrong: 52 Women Who Changed Science–And the World*. New York: Broadway Books.

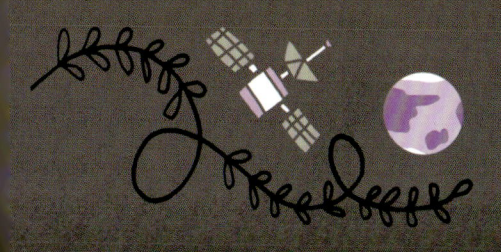

ACKNOWLEDGMENTS

I first want to thank all of the women who are working in science now. Through their passion and hard work they are creating a better future. And, of course, thank you to the women who are staying up way too late studying and researching to become the best doctors, scientists, and engineers they can be. I also want to thank all the young girls who are playing with bugs, looking at the stars, and driving their parents nuts by taking apart old machines.

Special thanks to Thomas Mason IV for all of his love, support, amazing suggestions, and bagels while I was creating this book. Thanks to Mia Mercado for all her grammar know-how. Another very special thank you for Aditya Voleti for helping me understand all of the math in this book, and for his great suggestions, expert grammar skills, help with fact checking, and of course his delicious biryani.

A special thank you to my editor, Kaitlin Ketchum, book designers Angelina Cheney and Tatiana Pavlova, and the rest of the talented publishing team at Ten Speed Press for all of their hard work and expertise! And lastly, a big thank you to my literary agent Monica Odom for finding my work and believing in me.

ABOUT THE AUTHOR

Rachel Ignotofsky grew up in New Jersey on a healthy diet of cartoons and pudding. She graduated with honors from Tyler School of Art's graphic design program in 2011. Now she lives in beautiful Kansas City, Missouri, where she spends all day drawing and learning as much as she can. She has a passion for taking dense information and making it fun and accessible and is dedicated to creating educational works of art.

    Rachel is inspired by history and science and believes that illustration is a powerful tool that can make learning exciting. She uses her work to spread her message about education, scientific literacy, and powerful women. She hopes this book inspires girls and women to follow their passions and dreams.

    This is Rachel's first book and she plans on writing many more in the future. To see more of Rachel's educational art and learn more about her, please visit www.rachelignotofskydesign.com.

INDEX

A
Abraham, Karl, 23
Aiken, Howard, 56-57
Ammal, Janaki, 114
Anning, Mary, 14-15
Antheil, George, 69
Anthropology, 91
Astronomy, 7, 8-9, 12-13, 32, 50-51, 86-87, 115
Astrophysics, 50-51, 100-101
Ayrton, Hertha, 20-21
Ayrton, William, 21

B
Babbage, Charles, 17
Ball, Alice, 44-45
Barre-Sinoussi, Françoise, 114
Bascom, Florence, 26-27
Bath, Patricia, 96-97
Becquerel, Henri, 29
Berson, Solomon, 81
Biochemistry, 46-47, 64-65, 73
Blackburn, Elizabeth, 104-105
Blackwell, Elizabeth, 18-19
Blackwell, Emily, 19
Bodichon, Madame, 21
Botany, 30-31, 52-53, 114
Boulenger, George Albert, 49
Buck, Linda, 114
Burnell, Jocelyn Bell, 100-101
Byron, Lord, 17

C
Carson, Rachel, 58-59
Chase, Mary Agnes, 30-31
Chemistry, 7, 28-29, 32, 44-45, 78-79, 114, 115
Chimpanzees, 90-91
Civil Rights Act, 33
Clark, Kenneth, 71
Clark, Mamie Phipps, 70-71
Clarke, Edith, 40-41
Claytor, W. W. Schieffelin, 75
Coe, Ernest, 43
Cohen, Stanley, 63
Computer programming, 7, 16-17, 33, 56-57, 88-89
Conservation, 42-43, 58-59
Cori, Carl, 47
Cori, Gerty, 46-47
Crick, Francis, 7, 79
Crystallography, 64-65, 79, 115
Curie, Marie, 7, 28-29, 32, 114
Curie, Pierre, 29

D
Daly, Marie, 32
Douglas, Marjory Stoneman, 42-43

E
Earle, Sylvia, 92-93
Easley, Annie, 88-89
Einstein, Albert, 7, 35, 39
Electrical engineering, 20-21, 40-41
Elion, Gertrude, 72-73
Engineering, 20-21, 36-37, 40-41, 95, 115
Entomology, 10-11
Equal Pay Act, 33
Ergonomics, 36-37

F
Flanigen, Edith, 115
Ford, Kent, 87
Fossil collecting, 14-15, 115
Franklin, Rosalind, 7, 78-79
Freud, Sigmund, 23

G
Genetics, 6, 7, 24-25, 52-53, 98-99
Geology, 26-27, 106-7
Gilbreth, Frank, 37
Gilbreth, Lillian, 36-37
Goeppert-Mayer, Maria, 54-55
Goodall, Jane, 90-91
Greider, Carol, 105

H
Hahn, Otto, 35
Harrison, Anna Jane, 114
Herpetology, 48-49
Herschel, Caroline, 32
Hewish, Antony, 101
Hill, Ellsworth Jerome, 31
Hitchcock, Albert, 31
Hitchings, George, 73
Hodgkin, Dorothy, 64-65
Hopper, Grace, 7, 56-57
Horney, Karen, 22-23
Huckins, Olga, 59
Hypatia, 8-9, 33

I
Inventing, 20-21, 33, 40-41, 56-57, 68-69, 96-97

J
Jackson, Miles, 44
Jackson, Shirley Ann, 114
Jemison, Mae, 108-9
Johnson, Katherine, 74-75
Joliot-Curie, Irène, 114

K
Kovalevskaya, Sofia, 115
Krafft, Katia, 106-7
Krafft, Maurice, 106, 107

126

L
Lab tools, 60–61
Lamarr, Hedy, 68–69
Leakey, Louis, 91
Leakey, Mary, 115
Lederberg, Esther, 82–83
Lederberg, Joshua, 83
Levi-Montalcini, Rita, 62–63
Lovelace, Ada, 16–17

M
Mandl, Fritz, 69
Marine biology, 58–59, 92–93
Masters, Sybilla, 33
Mathematics, 8–9, 12–13, 17, 21, 33, 38–39, 41, 74–75, 89, 112–13, 115
Mayer, Joe, 55
Mayer, Louis, 69
McClintock, Barbara, 6, 52–53, 105
Mechanical engineering, 36–37
Medical physics, 81
Medicine, 18–19, 46–47, 62–63, 72–73, 76–77, 80–81, 96–99, 104–5, 108–11, 114
Meitner, Lise, 34–35
Merian, Maria Sibylla, 10–11
Microbiology, 82–83
Milbanke, Anne Isabella, 17
Mirzakhani, Maryam, 112–13
Mitchell, Maria, 115
Molecular biology, 104–5
Morgan, Thomas Hunt, 24, 25
Moser, Edvard, 111
Moser, May-Britt, 110–11

N
Neurology, 62–63
Neuroscience, 110–11
Noether, Emmy, 7, 38–39
Nüsslein-Volhard, Christiane, 98–99

O
Oncology, 76–77
Ophthalmology, 96–97

P
Paleontology, 14–15
Particle physics, 102–3
Payne-Gaposchkin, Cecilia, 7, 50–51
Pharmacology, 72–73
Physics, 7, 28–29, 34–35, 38–39, 54–55, 66–67, 74–75, 114. *See also* Astrophysics; Medical physics; Particle physics
Piscopia, Elena, 33
Primatology, 90–91
Procter, Joan Beauchamp, 48–49
Psychology, 22–23, 36–37, 70–71, 110–11

Q
Qian Yiji, 13

R
Ride, Sally, 115
Rocket science, 88–89
Roebling, Emily, 115
Rubin, Vera, 86–87
Russell, Henry, 51

S
Scientific illustration, 10–11, 31
Space travel, 32, 94–95, 108–9, 115
STEM fields
 definition of, 7
 statistics on women in, 84–85
Stevens, Nettie, 24–25
Swain, Sandra, 76

T
Tereshkova, Valentina, 32, 94–95
Thomas, Tessy, 115
Ting, Samuel, 103

V
Virology, 114
Volcanology, 106–7

W
Wang Zhenyi, 12–13
Watson, James, 7, 79
Wilson, Edmund, 25
Wright, Jane Cooke, 76–77
Wu, Chien-Shiung, 66–67
Wu, Sau Lan, 102–3

Y
Yalow, Rosalyn, 80–81
Yonath, Ada, 115

Z
Zakrzewska, Marie, 19
Zoology, 48–49, 59
Zwicky, Fritz, 87

DEDICATED TO MY MOM AND DAD.

COPYRIGHT © 2016 BY RACHEL IGNOTOFSKY

ALL RIGHTS RESERVED.
PUBLISHED IN THE UNITED STATES BY TEN SPEED PRESS, AN IMPRINT OF THE CROWN PUBLISHING GROUP, A DIVISION OF PENGUIN RANDOM HOUSE LLC, NEW YORK.
www.crownpublishing.com
www.tenspeed.com

TEN SPEED PRESS AND THE TEN SPEED PRESS COLOPHON ARE REGISTERED TRADEMARKS OF PENGUIN RANDOM HOUSE LLC.

A FEW OF THE ILLUSTRATIONS IN THIS WORK FIRST APPEARED IN SLIGHTLY DIFFERENT FORM ON VARIOUS ONLINE PLATFORMS.

LIBRARY OF CONGRESS CATALOGING-IN-PUBLICATION DATA IS ON FILE WITH THE PUBLISHER.

HARDCOVER ISBN: 978-1-60774-976-9
eBOOK ISBN: 978-1-60774-977-6

PRINTED IN CHINA

DESIGN BY TATIANA PAVLOVA AND ANGELINA CHENEY

16 17 18 19

FIRST EDITION